Linguistische
Arbeiten 61

Herausgegeben von Herbert E. Brekle, Hans Jürgen Heringer,
Christian Rohrer, Heinz Vater und Otmar Werner

Wortstellung und Bedeutung

Akten des 12. Linguistischen Kolloquiums
Pavia 1977
Band 1

Herausgegeben von
Maria-Elisabeth Conte,
Anna Giacalone Ramat und Paolo Ramat

Max Niemeyer Verlag
Tübingen 1978

CIP-Kurztitelaufnahme der Deutschen Bibliothek

Linguistisches Kolloquium ⟨12, 1977, Pavia⟩
Akten des 12. [Zwölften] Linguistischen Kolloquiums : Pavia 1977 / hrsg. von Maria-Elisabeth Conte . . . – Tübingen : Niemeyer.
(Linguistische Arbeiten ; . . .)

NE: Conte, Maria-Elisabeth [Hrsg.]

Bd. 1 → Wortstellung und Bedeutung

Wortstellung und Bedeutung / hrsg. von Maria-Elisabeth Conte . . . – 1. Aufl. – Tübingen : Niemeyer, 1978.
(Akten des Zwölften Linguistischen Kolloquiums ; Bd. 1) (Linguistische Arbeiten ; 61)
ISBN 3-484-10303-5

NE: Conte, Maria-Elisabeth [Hrsg.]

ISBN 3-484-10303-5

INHALTSVERZEICHNIS

INHALTSVERZEICHNIS ZU BAND 2

VORWORT

Das 12. Linguistische Kolloquium fand vom 6.-9. September 1977 in Pavia (Italien) statt und bot die Gelegenheit zu intensiveren Kontakten zwischen italienischen Linguisten und Linguisten aus anderen europäischen Ländern, insbesondere aus der Bundesrepublik Deutschland.

Ebenso wie die Akten aus Tübingen (10. Linguistisches Kolloquium) und Aachen (11. Linguistisches Kolloquium) erscheinen auch die Akten von Pavia in zwei Bänden. Aus der grossen Zahl der Referate sollte man jedoch nicht den Fehlschluss ziehen, es hätte sich um ein grosses Kolloquium gehandelt. Ganz im Gegenteil: die Zahl der Teilnehmer war relativ niedrig. Nur siebzig Linguisten waren gekommen, die aber fast alle ein Referat zur Diskussion stellten. Es war also kein Kongress, sondern ein wirkliches Kolloquium, das in Stil und Atmosphäre wieder mehr den Linguistischen Kolloquien der ersten Jahre glich.

Wie es bei den Linguistischen Kolloquien üblich ist, gab es auch in Pavia keinerlei thematische Einschränkung für die Referate. Als Organisatoren des Kolloquiums hatten wir aber schon im 1. Rundschreiben dazu eingeladen, Referate zur Frage der Wortstellung vorzubereiten. Zu diesem Thema wurden acht Referate vorgetragen, die in zwei Sitzungen intensiv diskutiert wurden. Das Problem der Wortstellung bildete somit einen Schwerpunkt des Kolloquiums, der sich jetzt im Titel des 1. Bandes (Wortstellung und Bedeutung) niederschlägt.

Wir haben auch die anderen Referate einigen grob festgelegten thematischen Bereichen zugeordnet. Diese Zuordnung war häufig problematisch, und einige Referate hängen nur sehr locker mit dem Thema der Sektion, in die sie eingegliedert wurden, zusammen. Im 1. Band folgen auf die erste Sektion: Wortstellung und Typologie drei weitere Sektionen: Syntax und generative Theorie; Semantik; Automatische Übersetzung. Der 2. Band (Sprache im Kontext) enthält die Sektionen: Soziolinguistik; Argumentation und Kommunikation; Interferenzprobleme; Sprechakte und Texte; Stil und Mythos.

Die beiden Bände bilden eine Einheit und dokumentieren bestimmte Forschungsinteressen, die augenblicklich in Europa von ganz jungen und nicht mehr ganz so jungen Linguisten vertreten werden.

Was die technische Seite der Herausgabe der Akten betrifft, so richten sich die Bände nach dem Muster der Akten von Tübingen und Aachen. Die Autoren haben ihre Beiträge selbst nach einheitlichen Schreibanweisungen auf Spezialpapier geschrieben, und die Beiträge sind in dieser Form vom Verlag gedruckt worden. Die Verantwortung auch für die Form lag voll bei den Autoren. Wir haben als Herausgeber geringfügige Fehler selbst korrigiert, einige Beiträge, falls die äussere Form es verlangte, haben wir an die Autoren zurückgeschickt und neu schreiben lassen.

Nicht alle Beiträge entsprechen voll den typographischen Normen. Aber bei der Wahl zwischen formaler Homogenität der Beiträge und einem raschen Erscheinen der Bände haben wir der rechtzeitigen Veröffentlichung den Vorrang gegeben.

Zum Schluss möchten wir all denen unseren Dank aussprechen, die zum Gelingen der Tagung in Pavia beigetragen haben: den Organisatoren des 11. Linguistischen Kolloquiums in Aachen, die uns in den verschiedenen Phasen der Organisation unseres 12. Linguistischen Kolloquiums berieten; dem Rektor der Universität Pavia, der die Tagungsräume zur Verfügung stellte und zur Eröffnung des Kolloquiums einen Aperitif anbot; den Direktoren der collegi universitarî, die angenehme und billige Unterkünfte für die Teilnehmer bereitstellten; der Stadt Pavia, die den Ausflug zur Kartause von Pavia und nach Vigevano ermöglichte; nicht zuletzt Herrn G.Gilberti, der mit viel Umsicht vor,während und nach der Tagung das Sekretariat des Kolloquiums leitete.

Pavia, im Januar 1978

Maria-Elisabeth Conte
Anna Giacalone Ramat
Paolo Ramat

1. WORTSTELLUNG UND TYPOLOGIE

TOWARDS A TYPOLOGY OF THEME AND RHEME: SOV LANGUAGES

László Dezső

I shall discuss the following topics in this paper:

1. the place of the process of theme and rheme in syntactic typology,
2. the typological process of theme and rheme in SOV languages,
3. the change of the rules of this process in connection with the change of SOV type.

It goes without saying that I can only comment on the main issues of these topics in this brief paper.

1. The place of the process of theme and rheme in syntactic typology

In order to clarify this problem, I must /a/ interpret the notion 'process', /b/ mention those processes that are closely connected with that of theme and rheme, and /c/ to describe theme and rheme as a process in its relation to other processes.

The use of the term 'process' may be new in the metalanguage of typology but what is meant by it is the description of a typological subsystem by means of rules and principles. It assumes that a subsystem of language can be described in a synthetic approach starting from an invariant base and resulting in typological structures.

An invariant base structure is characterized in the universal terms of typology. The derivation of resulting structures is given by a calculus, i.e. by a logically possible set of derivational rules. Fact-finding typology is interested in those of them which result in various types of derived structures, but the non-productive rules are also relevant to a subsystem theory. Principles that underly and explain other rules are of particular importance. The resulting structures are characterized also by formal means expressing them. What I have said is not novel from the point of view of the methodology of linguistics. It reflects the views of the Leningrad group of typology and is close to the approach of the Cologne project on language universals, though the latter uses the term 'dimension'.

I do not intend to enumerate the processes more or less adequately studied by typology. I call the attention only to two of them: structuring of a sentence and diathesis. Instead of describing them in any detail I shall regard them from the point of view of theme and rheme.

I shall use the term 'functional sentence perspective' in order to cover the whole communicative organization of the sentence that includes both diathesis, i.e. the derivation of active, passive, impersonal etc. sentences, and their further variants resulting from the application of the rules of word order, sentence stress and morphological means expressing theme and rheme. Thus, functional sentence perspective is a common term for both diathesis and theme-rheme rules. It consists of two cycles: first, the various diatheses are derived, then the theme and rheme variants of each diathesis are formed.

In order to derive the basic /active/ and then the secondary /passive, impersonal/ diatheses and to form the various types of sentence structures, invariant structures are needed. They consists of a predicate, of obligatory arguments or <u>actants</u>, of free arguments or <u>circumstants</u>. The obligatory arguments may have features or labels reflecting the active or non-active role of the argument in an action, state or event. They depend on the features or classes of the predicate. The free arguments denote the place, time, manner of an action, state or event.

In the basic diathesis, in its basic theme-rheme variant, one of the obligatory arguments is the subject or the main element of the theme, another argument is the object, or the main element of the rheme. In the basic diathesis the arguments take the role of subject and that of object according to their degree of "communicative dynamism", i.e. according to a universal scale of value for communication: the future subject has a lesser, the future object has a higher value for communication.

There are two principles of morphological marking: subject marking and object marking. The means of marking are the morphemes of the noun or those of the verb. There are two types of subject marking languages: ergative and active. In ergative languages only a subject of a transitive verbal predicate is marked, in active languages the subject of an intransitive predicate is also marked. The object is unmarked in both types. In object marking or nominative languages, the subject is unmarked, the object is marked. In marking the subject or object, the features or classes of predicate are reflected /KLIMOV 1973, DEZSŐ 1978/. There are languages neutral from the point of view of marking the subject or the object.

These types of sentence structure may be derived by a calculus proposed by Kibrik /KIBRIK 1978/. The subtypes show the influence of determination and aspect, e.g. in the partitive subtype of nominative languages, the nominative, accusative vs. partitive affixes are chosen according to the definiteness of the subject, of the object and the perfective or non-perfective aspectuality of the verbal predicate /DEZSŐ 1978/.

These sentence types have or may have their secondary diathesis variants: passive, impersonal etc. sentence structures. The calculus of diathesis was described by Khrakovsky and its typology was elaborated by the Leningrad group /TIPOLOGIJA 1974, CHRAKOVSKIJ 1975/. The process of theme and rheme starts from the variants of diathesis. The types of the latter are the invariant structures for the variants of theme and rheme.

2. The process of theme and rheme in SOV languages

The process of theme and rheme has as its marked role the rheme The means of its marking are sentence stress, word order and special morphemes. Sentences stress is universal, word order has various types, the use of morphological means is limited and it will not be examined here. A theme-rheme theory explaining this process differentiates various kinds of rheme: focus assuming no previous context, emphasis, when a piece of information is lacking, and contrast, when a previous contrasting context is assumed.

Typology has studied the theme-rheme problems of the basic diathesis of nominative sentences, consisting of a verbal predicate, subject and direct object. The terms subject, direct object and verbal predicate denote the parts of a nominative sentence, hence "surface" categories. These results can be extrapolated to ergative, active and neutral languages, but then the terms must be reconsidered. Typology disposes a considerable amount of non-systematized data concerning sentences with an indirect object and with various adverbials.

The basic theme and rheme variant of basic diathesis may be one of the six possible combinations of S, O and V. The basic variant does not require a special previous context, and therefore its rheme is only focused. The data of various languages show that 4 of the six variants are used as basic ones. In three of them: SOV, SVO and VSO, the subject precedes the object and the degree of communicative dynamism increases in the linear order but in VOS type it decreases. In all of the possible sentences, stress marking the rheme is posited on the O /see DEZSŐ 1978, DEZSŐ-SZÉPE 1974/.

I shall confine myself to the SOV type, confronting it with the SVO type.

In the secondary variants, the order of O and S is the reverse because its use assumes a previous context containing O. The primary place of stress marking the rheme is on the subject:

/1/	SOV	/2/	SVO	/3/	VSO
	OSV		OVS		VOS or SVO

In the primary and secondary variants of types SOV and SVO, the place of verb is the same in both variants. In a tertiary variant of both types the verb is in a sentence initial position and followed by the subject and object: VSO. The sentence stress is on the verb in SOV languages, it is on S in SVO languages. /The latter may stress V, but this may be a relic of an earlier stage of SOV type./

The place of an indirect object and that of various adverbials is determined by the universal principle of semantic cohesion: the direct object is usually the closest to the verb, the indirect object stands further. The place of the adverbials depends on their relation to the predicate: manner adverbials tend to be close to the verb, most of the adverbials are further from the verb than the objects and setting adverbials may be in sentence initial position. This is the usual position of the various parts of a sentence:

/1/ $A^{set}SAO^{id}O^{d}V$ /2/ $A^{set}SVO^{d}O^{id}A$

The most important word order variants are usually described in the studies of individual languages and typology could formulate their regularities. Very little is known about the principles of sentence stress. What I am going to say should be regarded only as tentative and temporary statements based on the data of Uralic, Altaic and some Indo-European languages /cf. BASKAKOV 1975, ČERNYŠEV 1965, GAZIZOV 1966, IVANOV 1963, MUSAJEV 1961, NUŠAROV 1974, SOVREMENNYJ/.

The usual place of sentence stress and hence that of rheme is the position immediately preceding the verb. If this position is occupied by an element that cannot be moved from this place /e.g. an indirect object in absolute case in Turcic languages/, an element further before the verb can be stressed. If X denotes a stressed element, Y an unstressed one, then:

/i/ XV /ii/ XYV

demonstrate the variants of the preverbal stressing principle. In languages having no elements with fixed preverbal position /e.g. in Hungarian/ only variant /i/ is possible, i.e. a stressed element must immediately precede the verb. In languages with both variants, a stressed element may be separated from the verb under special conditions.

In rigid SOV languages, no element can follow the verb. If a postverbal position is admitted, the element following the verb is either unstressed or has an emphatic or contrastive stress denoted by X:

/iii/ VY /iv/ VX

In SVO languages, the usual place of the focused rheme is after the verb either in an immediately postverbal position of after an unstressed element:

/i/ VX /ii/ VYX

In a preverbal position the rheme expresses emphasis or contrast in SVO languages with free word order:

/iii/ XY

Thus, the principles of stressing the rheme are the opposite in SOV and SVO languages. The rules of individual languages are more complicated because they are often of mixed type. If in SOV languages an element does not stand close to the verb, as a result of the semantic cohesion principle, but is stressed, it moves to an immediately preverbial position, if it can do so, and instead of $SAO^{id}O^{d}V$ one finds:

$SAO^{d}O^{id}V$ or $SO^{id}O^{d}AV$ or $AO^{id}O^{d}SV$ etc.,

where O^{id}, A, S are stressed. /Of course, the position of unstressed nominal elements may vary./ If the immediately preverbal position is occupied, the stressed element stands before the verb and tends to the closest possible position; e.g. $SO^{id}AO^{d}V$ or $AO^{id}SO^{d}V$ etc.

The verb may be stressed in any position, but if it is in sentence initial position, it has an obligatory emphasis or contrast.

3. The change of theme-rheme rules in connection with the change of SOV type

If a language changes its type of word order from SOV to SVO, it has to change the principles of sentence stress as well. The restricted subtypes of both types have no common variant:

/1/	SOV	/2/	SVO
	OSV		OVS

and no change of type is possible. The existence of a common variant VSO is of no relevance. A change of type is possible in cases of free word order, when the basic order and the stressing principles of the two types differ but the untypical variants are also used:

/1/	SVO	/2/	SOV
	OVS		OSV

The untypical variants of SOV languages has one /or, in cases of 4 elements, several/ elements in post-verbal position. They may be either unstressed or contrasted but they cannot have simple focus.

An SOV language must change its stressing principles to those of SVO:

	SOV:	/i/ XV	/ii/ XYV	/iii/ $V\underline{X}$ or VY
to	SVO:	/i/ VX	/ii/ VYX	/iii/ $\underline{X}V$

The postverbal position must also be a focused one instead of an unstressed or in addition to a contrasted position of SOV type. The preverbal position cannot be used in cases of focus, but it will preserve its emphatic or contrastive status /that was not specially mentioned earlier/.

The main principle of preverbal stressing changes to a postverbal one. Thus one arrives at the principles of Russian /cf. DEZSŐ 1978/. Proto-Indo-European and Proto-Slavic were SOV languages, Modern Russian belongs to SVO languages with free word order. It has preserved the stress on the sentence initial verbs: $\underline{V}$SO instead of V$\underline{S}$O with a stressed S in such an originally SVO language, like Swahili.

The process of transition from an SOV type to SVO type is more complicated, of course. Many languages have variants of mixed type in cases of four elements, e.g. SO^{d}VO^{id} with a direct object before and with an indirect one after the verb. Other languages have double word order depending on the structure of the sentence. Both phenomena can be found in Hungarian /cf. DEZSŐ 1978/. Double word order is a result of change and is connected with differences in determination and aspect in Hungarian and in other languages. There are other variants of double word order, e.g. sentences with synthetic and analytic verbal predicates and the difference between main and subordinate clauses in German /cf. RAMAT/.

LITERATURE

BASKAKOV, N.A. /1975/: Istoriko-tipologičeskaja charakteristika struktury tjurskich jazykov. Moskva: Nauka.

ČERNYŠEV, V.A. /1965/: Sintaksis prostogo predloženija v chindi. Moskva: Nauka.

CHRAKOVSKIJ , V.S. /1975/: Isčislenije diatez. Diatezy i zalogi. Leningrad.

DEZSŐ, L. /1978/: Studies in syntactic typology and contrastive grammar. Budapest: Akadémiai Kiadó.

DEZSŐ, L. - SZÉPE, Gy. /1974/: Contribution to the topic-comment problem. Topic and comment, contextual boundness and focus. /ed. Ö. Dahl/ Hamburg, 69-93.

GAZIZOV, R.C. /1966/: Sopostavitel'naja grammatika tatarskogo i russkogo jazykov. Kazan'.

IVANOV, Vjač. V. /1963/: Chettskij jazyk. Moskva: Nauka.

KIBRIK, A.Je., Ergativity standard and Daghestan languages: Ergative syntax /ed. F. Plank/ /to appear in 1978/.

KLIMOV, G.A. /1973/ Očerk obščej teorii ergativnosti. Moskva: Nauka.

MUSAJEV, D.I. /1961/: Porjadok slov v predloženijach anglijskogo i azerbajdžanskogo jazykov. Baku.

NUŠAROV, M.M. /1974/: Ramočnaja konstrukcija v nemeckom jazyke i zamykanije v uzbeksom. Samarkand.

RAMAT, P. /1976/: Ist das germanische ein SOV-Sprache? Akten des V. Internationalen Germanisten-Kongresses. Cambridge. 1975: 25-35.

Sovremennyj kazachskij jazyk. /1962/ Alma Ata.

Tipologija passivnych konstrukcij. /1964/ Leningrad: Nauka.

PROBLEMS WITH OV/VO WORD ORDER

Klaus-Peter Lange

1. This paper deals with a theory of word order which can be roughly defined by citing the names of Greenberg, Lehmann, and Vennemann. Their theory also claims to be typological. I will firstly give some reflections on the untenability of their approach, and secondly suggest a new way out of the difficulties into which it is bound to run.

2. Although I believe that this theory of word order is widely known, it is necessary for my purposes to recall some notions that are basic to it. VENNEMANN (1973a), (1973b), (1974), (1975) subscribes to GREENBERG's (1961) idea that there are, on a language universal scale, three basic types of word order, viz. the VSO, the SVO, and the SOV type, which mean that the type of a language is determined by the relative order of subject, object, and verb. But more than this, the relative order of the O and V constituent is supposed not only to reflect the relative position of verb and object, but also that of other constituents which stand in the same relation of modified and modifying element to each other, as verb and object do. These other relations are for instance noun/qualifying adjective or noun/noun in the genitive case.

Since, however, the majority of languages, as for example English or German, does not fit into this very rigid pattern, as Greenberg himself admitted freely from the very beginning (German has OV order in subordinate clauses and can have VO in main clauses), Greenberg had to restrict his material, and he did so by considering nothing but the relative order of S, O, and V in independent declarative clauses with nominal subject and object. But Greenberg's interest was a statistical one, and however he would have restricted his material he would have reached undebatable results.

Turning now to Vennemann, we witness the attempt to get all those conflicting facts into line which Greenberg had put into small statistical rests. Obviously driven by a sense of purism and idealism, Vennemann tries to make a rule out of Greenberg's statistical main figures. On the basis of the fact that most of the languages with OV word order put modifying constituents before modified ones, and most of the VO languages reverse this order, he sets up his "principle of natural serialization" which says that OV languages tend to serialize modifiers (in his terms "operators") before their respective modified elements (in his terms "operands"), whereas it is the other way around in VO languages.

But how does he cope with the undenyable fact that languages in very rare cases only, like Arabic or Japanese, give evidence for the correctness of the principle of natural serialization? The answer is to be found in VENNEMANN (1974: 347):

> ... if it is so 'natural' to have a certain word order or one of two kinds of word order, why are not all languages consistent in the word order type that they represent? The answer is, of course, that languages change.

If I interprete Vennemann's interest in historical word order change correctly, it is no interest in itself, rather, he has come to this field of research through the necessity of remediating, or more unpolitely of concealing, the inadequacies of his "natural serialization principle". It is this principle which my further considerations will now center on. I will not go into a discussion of Vennemann's theory of typological change. This has for instance been done by LI/THOMPSON (1974) and KLEIN (1975).

But it could be worthwile to consider the natural serialization principle more closely since it is (though not under this special name) also present in the work of LEHMANN (1974), more correctly: Lehmann invented what Vennemann later called the natural serialization principle. While Greenberg believed that there is a threefold basic classification of languages as SOV, SVO, and VSO, Lehmann thought it more adequate to postulate a double basis for the typology of languages. The reason was that he believed there to be a mechanical correlation between the relative position of verb and object and the relative position of those other operators and operands I mentioned, i.e., when a language changes its basic verb position, the correlated phenomena will also change subsequently. He does not say why this should be so, why changing the relative position of the verb should result in a subsequent order change of other operator/operand relationships. Vennemann, who adopts Lehmann's fundamental views, also fails to give an explanation for the structurally leading role of the verb.

To determine why both linguists gave the verb the prominent role of the heading function in typology, we are thrown upon guess work. A solution could be that in the generative process the verb is the most fundamental operand since it is generated earlier than all the other possible operands, at least in a case grammar which LEHMANN (1974) favours. Another problem is the mechanicity of word order change with the verb change as its stimulus and the other changes as responses. Why should there be such a mechanical law in language life?

2.1. I now turn to a special argument against taking the position of the verb as basic for typological considerations. First, I will define what I think to be autonomous syntax. It is a linguistic conception which views all syntactic processes (I mean this in a generative sense) as independent of any kind of meaning functions. The autonomous syntactician believes that the arrangement of words in a sentence depends on psycholinguistic facts which concern only the arrangement of words, and that the possibilities of arrangement within the human mind are preformed, i.e. innate. In any case, the human mind does not have too much freedom in putting words together to form sentences. Thus, word order patterns are not only preformed on a deep generative level but also on the surface and on the intervening levels.

Looking back to the typological theory we have been discussing, we notice that it fits into the general conception of autonomous syntax. VSO, SOV, and SVO, or OV vs. VO are syntactic patterns which seem to belong to a special linguistic property of the mind, more exactly to the *syntactic* property of the mind, which is not capable of avoiding the production of those patterns in concrete sentences.

An alternative conception of syntax, which was once proposed in the history of generative grammar but soon forgotten, would be the claim that the order of words is not guided by a purely syntactic mental capacity but by a semiotic capacity which controls word order with regard to meaning. It was Chomsky himself (now champion of autonomous syntax) who explicated this conception to a certain degree in his first book from 1957. Since knowledge of this early work of his has almost come down to zero by now, it is necessary to recall that it regarded transformations as meaning dependent, a view which was abandoned in CHOMSKY (1965), which initiated autonomous syntax.

I think it would be interesting and rewarding to go back to the Aspects book to look for the reasons why Chomsky came to favour autonomous syntax. Unfortunately I have no space to do so here. But he who does will come to the conclusion that his reasons for throwing meaning dependence of transformations overboard are extremely weak. His main reason was to make the overall theory simpler by having only deep structure as input to semantic interpretation. He achieved this alleged simplification at the cost of the invention of a new mental capacity which organizes word order without respect to meaning. This mysterious mental capacity had not existed in linguistic or psychological theory up to then.

Looking back now to CHOMSKY (1957) which regarded transformations as dependent on meaning, i.e., as meaning creating, I want to call this conception "semiotic syntax". It stands in contradiction to autonomous syntax since the latter depends on, and semiotic syntax rejects, the assumption of a purely syntactic mental capacity. In this respect autonomous syntax is consistent with both Chomsky's interpretive semantics and a variety of generative semantics models, Vennemann's model being only one of them. Chomsky has proposed a grammatical theory in which an autonomous syntactic component is primary input to secondary semantic interpretation. All those generative semantics theories (those of Lakoff, McCawley, Fillmore, and BARTSCH/VENNEMANN (1972)) make a semantic component primary, formulated in a somewhat modified predicate calculus, with an autonomous syntactic component as output. In both of these grammar types, which have been relatively well elaborated in the past decade, syntax and semantics have nothing in common. They are obviously meant to be correlated to different parts of the mind, to a syntactic part and a semantic part respectively.

Semiotic syntax, however, is based on the assumption that there is no mental distinction between handling meanings and ordered or unordered sets of language signs. Transformations are regarded as meaning creating or meaning adding. To take an example, autonomous syntacticians like VENNEMANN (1973a), (1973b), BARTSCH/VENNEMANN (1972), or HUBER/KUMMER (1974) assume that there must be a basic word order in German which is either SVO or SOV (in Huber/Kummer's terms) or VO vs. OV (in Vennemann's terms), and he who tries to write a grammar for German will find the basic word order by looking at independent declarative clauses, as Vennemann suggests. Huber/Kummer propose to make that word order basic which provides the possibility of deriving all deviant word orders in the easiest way. Both suggestions are unsemantic.

It is widely known that the position of the finite verb is a central issue of German grammar. There are mainly three places in the German sentence for the verb to stand: initial, second, and final position. The autonomous syntactician, who believes that one of these positions only is the basic or underlying one, chooses among them in building up the syntactic component of German grammar, and he does so without respect to a difference in meaning expressed by the different positions of the verb.

The semiotic syntactician, however, will not have to choose among three verb positions as basic, for he will inquire: what is the meaning difference expressed by the varying positions of the verb? He will notice that there is such a difference with respect to the illocutionary force of the clause. Take for example the following German sentence:

(1) Ob er noch kommt? (= I wonder whether he's still coming)

We can analyse (1) in the subsequent manner:

(2) ob (er noch kommt)

Here we have the proposition of (1) in parentheses and the indicator of illocutionary force as its function. The proposition itself shows the verb <u>kommt</u> in final position. The illocution indicator <u>ob</u> is a sign by its own, which takes the propositional content of the sentence to form a kind of yes/no question. We can replace the illocution indicator <u>ob</u> by other signs of this type, as for example daß, without changing the propositional content. By this we will receive the independent grammatical sentence:

(3) Daß er noch kommt! (= I'm surprised that he's still coming)

This sentence adds to the proposition er noch kommt the illocutionary force of the speaker's surprise about what is said in the proposition. Illocutionary forces are of course not always expressed by independent signs, like ob or daß (in many cases in addition to illocutionary verbs like versprech-, erwart-, glaub- etc.). Very often, they are expressed by transformations involving the finite verb. The illocutionary meaning component of plain yes/no questions normally is - at least in German and English - denoted by a verb shift which transports the finite verb from its propositional position to a position of the propositional rest. Taking the parenthesized proposition of (2) as an example, we receive the correct yes/no question (4) by applying the verb shift transformation just described:

(4) Kommt er noch? (= Will he still come?)

An additional transformation filling the slot before the shifted verb will add another illocutionary force to the propositional content of (2), viz. the force of declaration:

(5)(a) Er kommt noch (= He's still coming)
(b) Noch kommt er (= At this moment, he's still coming)

Thus, the sign for the illocutionary force of yes/no questions is a simple shift transformation, whereas the sign for declarations or statements is a double shift.

Two remarks at least are necessary here: first, that the commonplace conception of the linguistic sign will have to be revised under this theory of semiotic syntax. Not only independent, physically representable, and regularly shaped strings of sounds are linguistic signs but also transformational processes like those I have described here. Second, I am applying a notion of illocutionary force which diverges somewhat from that in speech act theory, which does not make illocutionary forces dependent on special signs signalling some specific speech act. Semiotic syntax, however, is forced to identify signs (in the sense in which I am using this term here) with certain definite meanings one class of which are illocutions. Consequently, the transformational sign of declaration always has the meaning of a declaration. This is not contrary to the fact that declarations sometimes can be utilized as requests or commands in certain pragmatical circumstances.

It has by now become obvious that, in the framework of semiotic syntax, considerations of basic word order gain a new dimension. To ask what word order is basic or underlying is to ask what the order of constituents in the proposition is like. This order can be distorted by additional semiotic operations. It turns out to be arbitrary, under the present conception of semiotic syntax, to take the order of independent declarative clauses as basic.

German is OV on the propositional level, for here it serializes objects before the verb. On the higher level "proposition + illocution", German is everything: VSO, VOS, SVO, and OVS. This shows that these typological categories can no longer play any part in semiotic syntax. But the fact that German serializes objects before verbs in the proposition does not at all mean that German is a OV language in Vennemann's sense, for the natural serialization principle can, I think, be discarded and replaced by a much stronger principle of serializing operators with respect to their operands. It is this stronger principle I am going to deal with in the last section of this paper.

3. The question to be answered here is: what determines the actual position of operators relative to their operands on the propositional level? It is not answerable on the illocutionary level because here we can find a distortion of operator/operand relationships effected by element shift. There is an even higher level of semiotic processes (that of functional sentence perspective) which also effects a rearrangement of the sentence, and which I do not intend to go into here any further. The problem is how to determine the operator/operand relationships in terms of word order rules. Consider the following German noun phrases:

(6)(a) die heut-ig-e Jugend
(b) die Jugend von heute (= the youth of today)
(c) die Jugend heute

Everybody who knows German will admit that these NPs are equivalent in meaning apart from some slightly different stylistic flavor. Leaving aside the article die, we must, in each case, identify Jugend as operand and heut- as operator since the latter modifies the meaning of the former. In (6a), (6b) there are additional signs, among them the suffixed -e and the prefixed von, which serve to identify heut- as operator. They are operator indicators. Now we can tentatively put it as a rule that suffixed operator indicators require preposing of operators with respect to their operands, whereas prefixed operator indicators require postposing. Testing this assumption first with material from German, we notice that we have neither (7a) nor (7b):

(7)(a) *die Jugend heut-ig-e
(b) *die von heute Jugend

As to the English language, we can, for instance, regard the genitive -s as a suffixable operator indicator. According to our rule we will find (8a), but not (8b):

(8)(a) Peter's car
(b) *(the) car Peter's

Adverbs are operators with respect to VPs. If a language, like English, has developed a suffixable operator indicator for adverbs, it will put unstressed adverbs before the VP. So we will find sentences like (9a), whereas those like (9b) are ungrammatical:

(9)(a) Sam careful-ly sliced the salami
(b) *Sam sliced careful-ly the salami

As to Italian, we have the prefixed operator indicator di. Thus we will get (10a), but neither (10b) nor (10c):

(10)(a) il lago di Garda
(b) *il di Garda lago
(c) *di Garda il lago

It would be easy continuing the quotation of instances for this operator serializing rule from other languages. I could consider Turkish suffixes which necessitate preposing of operators, or I could mention the Persian ézâfé, which is a prefix with the consequence of postposing operators.

But what about operators without indicators? Here we discover first the possibility of integrating operators into the operand either in front or behind. Examples for front-integration are:

(11)(a) Hitler-Jugend (German)
(youth)
(b) igni- fer (Latin)
(fire) (bearing)
(c) polem-archos (Greek)
(war) (leader)
(d) fast-running

Examples for back-integration are:

(12)(a) Kong- en (Danish)
(king) (def. article)
(b) wa- toto (Swahili)
(classifier) (child)
(c) nach-Hause (German)
(to) (house)

The second possibility of handling operators without indicators is adposition with the same dichotomy of putting operators in front or behind. Front-adposition examples are:

(13)(a) (the) above address
(b) büyük köpek (Turkish)
(big) (dog)
(c) (een) boek lezen (Dutch)
(a) (book) (read)
(d) schnell fahren (German)
(fast) (drive)

As back-adposition examples we may take:

(14)(a) read a book
(b) leggere un giornale (Italian)
(read) (a newspaper)
(c) voiture rouge (French)
(car) (red)
(d) (die) Jugend heute (German)
(youth) (today)

Presented with actual language data, we come to a sixfold pattern of operator positioning on the propositional level, namely integration, adposition, and affixation each of which is to be divided into the subpossibilities of preposing and postposing. This framework is, I think, more adequate to treat the serialization of operators than Vennemann's principle of natural serialization because it enables us to describe word order as it really i s, not as it is evolving. By adopting this theory, which surely has to be elaborated, as I did in LANGE (to appear), we avoid a very serious consequence involved in Vennemann's principle, i.e. the consequence of dividing languages into good ones and bad ones, the good ones being those consistent with the principle, the bad ones being those not consistent (cf. BARTSCH/VENNEMANN (1972: 135)).

REFERENCES

BARTSCH, Renate / VENNEMANN, Theo (1972): Semantic structures. Frankfurt/M: Athenäum.

CHOMSKY, Noam (1957): Syntactic structures. The Hague: Mouton.

—— (1965): Aspects of the theory of syntax. Cambridge, Mass.: M.I.T. Press.

GREENBERG, Joseph H. (1961): "Some universals of grammar with particular reference to the order of meaningful elements". GREENBERG, J.H. (ed.), (1966): Universals of language. 2nd ed. Cambridge, Mass.: M.I.T. Press: 73-113.

HUBER, Walter / KUMMER, Werner (1974): Transformationelle Syntax des Deutschen I. Munich: Fink.

KLEIN, Wolfgang (1975): "Eine Theorie der Wortstellungsveränderung ...". Linguistische Berichte 37: 46-57.

LANGE, Klaus-Peter (to appear): Syntax und natürliche Semantik im Deutschen.

LEHMANN, Winfried P. (1974): Proto-Indo-European syntax. Austin: Univ. of Texas Press.

LI, Charles N. / THOMPSON, Sandra A. (1974): "An explanation of word order change SVO - SOV". Foundations of Language 12: 201-214.

VENNEMANN, Theo (1973a): Language type and word order. Trier: L.A.U.T.

—— (1973b): "Explanation in syntax". KIMBALL, John P. (ed.), (1973): Syntax and semantics. Vol. 2. New York: Seminar Press: 1-50.

—— (1974): "Topics, subjects, and word order: From SXV to SVX via TVX". ANDERSON, John / JONES, Charles (eds.), (1974): Historical linguistics. Amsterdam: North Holland: 339-376.

—— (1975): "An explanation of drift". LI, Charles N. (ed.), (1975): Word order and word order change. Austin: Univ. of Texas Press: 270-305.

ON SUBJECT: A CONTRIBUTION FROM CHILD ACQUISITION

Elisabetta Fava

The problem of defining grammatical notions such as Subject, Object and stating their possible ordering in natural languages has been an important issue in linguistic theory. However, in the transformational generative framework of the *Aspects* type[1], the redefinition of Grammatical Relations (GRs) in terms of constituent structure, largely defended by psycholinguists as a realistic model, diminished theoretical interest for such notions and discussions concentrated rather on the ordering of the elements in underlying structures. In fact, linearization, assigned to phrase structure rules of the base component to define the underlying level of the syntactic structures, implies that there is some single string ordering relation between any pair of elements introduced into any deep structure with no possibility of free word order at all. However, objections and criticisms were levelled both in a non generative framework, where the insufficiency of such definition of GRs was pointed out, and in a generative one, where proposals have been made to free syntax from underlyingly assigned precedence relations[2]. Recently, it has been largely stressed that the configurational definition of Subject is not possible on crosslinguistic bases: it has been proposed (COLE/SADOCK (eds.) 1977) that GRs in underlying free structures be defined in terms of basic properties and of the grammatical processes in which they are involved.

Operating within this last framework, I would specify those properties of GRs firstly acquired by the child in the form of a system of rules which conform to Italian and Veneto dialect. In particular, the relationship between the mastering of some of their basic coding properties (KEENAN 1976) and their linearization rule will be discussed. From the analysis it appears that while the agreement rules of Verb with the Subject are regularly mastered by the child in a very early stage, his order with respect to the Verb seems quite free or in any case the SVO order is not at all the prevailing one.

The study of the acquisition of GRs and word order has been firstly directed towards the demonstration of the innate character of syntax: researches by McNEILL (1966a, 1966b, 1970), SLOBIN (1966), ROEPER (1973a, 1973b) try to demonstrate that the existence of a preferential or fixed order can be explained only by referring to innate knowledge of syntax. McNeill, for instance, by assuming the configurational definition of Subject, Object as universal (McNEILL 1966a: 102) has tried to demonstrate the existence of basic language definitional universals, reflecting a specific linguistic ability and not necessarily a cognitive ability. The child's innate knowledge of GRs was reflected by the use of fixed order strategy even in languages which permit a relatively flexible ordering rule (children talk base strings directly even in Russian). The evidence of a fixed order strategy was reasserted by Roeper. She states that in German the subordinate clause strategy is a powerful mechanism for the discovery of the deep structure order. A child's awareness that the sentence was embedded and therefore had to be processed with ciclical transformations would entail an awareness that positions of (almost) all elements were fixed and therefore reflected deep structure. A serious challenge to such proposals based on purely syntactic strategy was given by BOWERMAN (1973a, 1973b), who showed that the evidence used to attribute to the child an understanding of basic GRs and the constituent structure they entail was inconclusive. On the contrary, there is evidence of variable word order from the time of the earliest two word utterances. She suggests that the word order is heavily influenced by input (adult's speech to child) and that children's initial combinations are based primarily on semantic considerations.

A partially similar position is encountered also in BATES (1976: 192-211), who, besides, stresses the importance of pragmatic factors in determining word order in the preoperational stage; also in Bates'work the interest for word order is strongly related to the verification of the existence of the notion of deep Subject. By analysing the statistical occurrences of the orders of Subject and Object with respect to the Verb and between them as appeared from the speech of two Italian children[3], she discovered a large use of orders different from the basic SVO order , i.e. orders that in Italian are marked or ungrammatical. In all cases observed there was a strong tendency to have semantic Subjects in final position. This fact was explained by Bates on pragmatic ground, confirming the hypothesis on child word order formulated in BARONI/FAVA/TIRONDOLA (1974). V a r i a t i o n s in the constituent order may be explained by the tendency to have, in a very early stage of acquisition, first new information and later given. In the same way,Bates argues, the tendency to have final Subject orders depends on the fact that semantic Subjects tend to be more likely old information. The mastering of Subject is relatively late for Bates, and coincides with a switch to have SVO order rigidly fixed, together with a switch from the use of proper name to express speaker and listener to the use of Subject Pronouns and, furthermore, with the productive marking of Subject/Verb agreement:

> These three standard developments - standard ordering, subject pronouns and subject verb agreement - can all be explained by the possibility that the child has just discovered the concept of syntactic subject. (BATES 1976: 209)

In this paper, I will argue that in Italian the linearization process of GRs towards a "preferential" SVO order is later with respect to the acquisition of such notions, whose basic coding properties are present at the beginning of the preoperational stage.

Our analysis is based on the tape recorded conversations collected from six children , following weekly their progress for a year and half, from 18 to 30 months of age. Recording of their verbalization was gathered at the nursery school where the children were used to live most of the time, while they were playing together or with the interviewer. Notes were made usually after each session and were based on the tape conversation and on the recollection of the situation at the time of the utterance. The whole corpus consists of about 2,800 utterances.

The background of these children is similar: the socioeconomic level of their families is working class: all the fathers are factory workers or small time tradesmen In four cases the mothers also are factory workers; in the other two cases they are housewives. Except for the two children whose mothers are housewives, all the others have been put in a nursery school before reaching seven months, and they spent most of the time there. Due to the poor ratio of staff to children and more generally of qualified teachers in this nursery school, the children were used to playing and interacting verbally most of the time with other children of the same age and it was only during the time they were home that they received specific attention.

The families of all children have been living for more generations in Valdagno, a small conservative town in Veneto (province of Vicenza) where dialect, a local variety of Veneto dialect, is still largely spoken by wide social strata. At home families generally speak dialect, even if the parents sometimes make efforts to speak Italian to the children. In the nursery school the children often prefer to talk with the others in dialect and with teachers in Italian. The interference between the two systems is often revealed. So these phenomena of interference, which mirror general problems of diglossia and bilingualism (MIONI/ARNUZZO 1977; MIONI/TRUMPER 1977), particularly problematic in the study of acquisition of the structure of a language, where they cannot be treated as separate systems nor in terms of deviations of one from the other (LABOV 1972), require an analysis in terms of a single continuum, which has switchings between Italian and Veneto.

In Italian and Veneto dialect the properties characterizing the Subject within a clause are represented in their formal structure by a clustering of rules that refer to underlying semantic features such as person/gender/number, rules generally called "agreement rules". This formal system appears to be controlled by the child from the beginning of his linguistic production in the process of mastering the suffixal Verb conjugation and the related pronominal system, that appear respectively in 30% (Verb) and in 5% (Pronoun) of his utterances.

The 85% of Verbs present either the finite Verb morphs or the past participle one (see table 1 for their respective curves), that must refer to person and number agreement with the Subject in the first case, or number and gender in the second. Pronouns appear firstly in the Subject case (full form) and later in the non-Subject case (clitic) (see table 2).

While the analysis of the agreement rules in the past participle is rather problematic since it is hard to decide what are the properties shared with Verbs or with Adjectives, it is mainly in the case of the finite Verb morphology that it is possible to concentrate on the systematically significant relations between Verb suffixes and pronominal system.

Table 1: percentages of Verbs categories on total Verbs

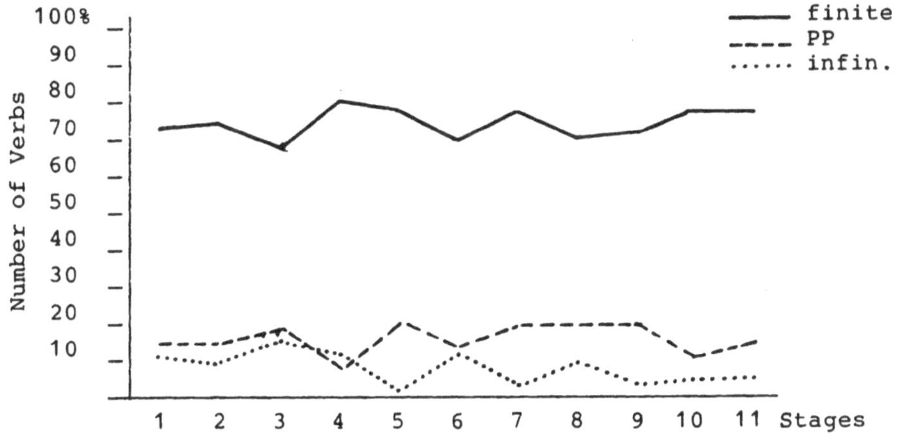

In fact, the child's ability to cover almost the whole range of morphological endings (see table 3) and a meaningful part of the pronoun system (see table 4) permits the verification of "a-greement" as a productive rule. This is not an accidental product of the fact that children use one unmarked form of the Verb to serve almost all functions.[4]

The appropriate use by the child of Verb morphology and Pronoun system is gleaned from a detailed analysis of the structure of children's discourse in the interaction with adults or with o-

Table 2: percentages of Pronouns

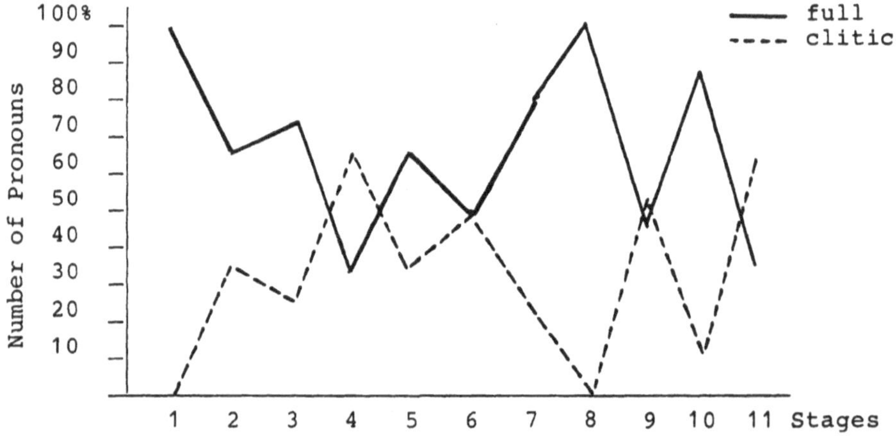

Table 3: percentages of finite Verbs

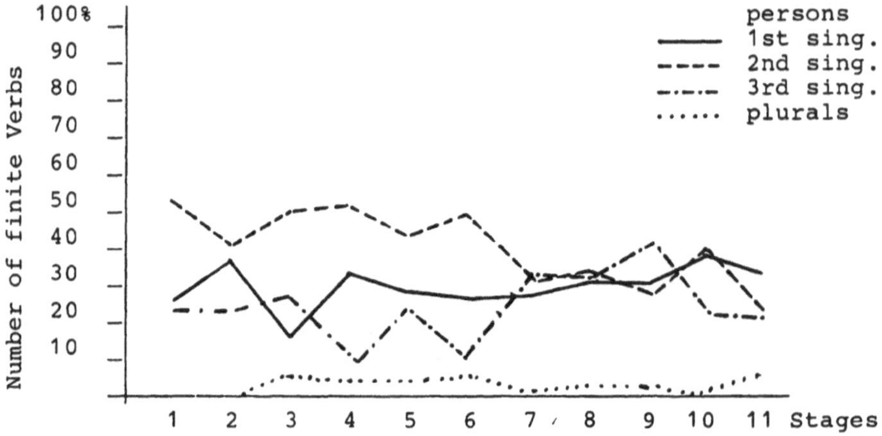

ther children. In the question/answer discourse type, for instance, it appears that in children's requests the hearer is always indicated by 2nd person Verb morphs and by 2nd person Pronoun.

(1) Massimo[5], 19 months of age (stage 2), is playing with little cars, talking alone. Realizing Interviewer's presence:
1.1. Massimo: ciao::/veni t(r)eno?
1.2. Int. : adesso non posso/
1.3. Massimo: (d)opo poi?

(2) Lele, 19 months of age (stage 2), with a puzzled look on his face:
.. ..
2.1. Lele: giochi tu?

Table 4: percentages of Subject Pronouns

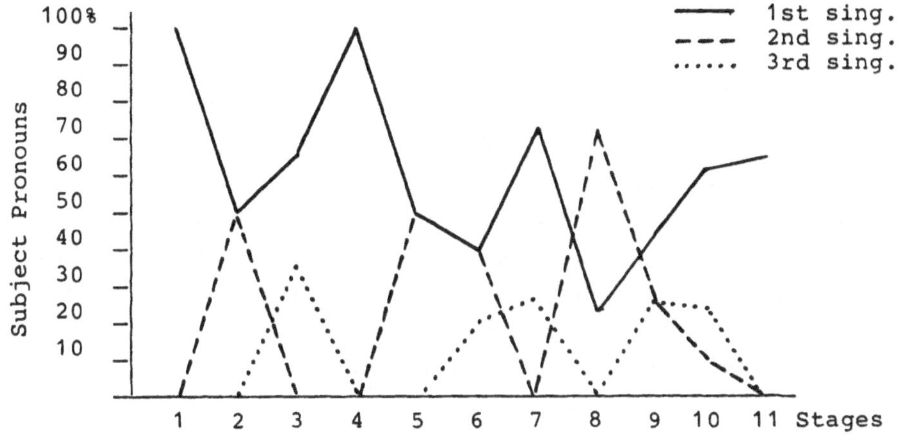

2.2. Int.: sì/se vuoi/
2.3. Lele: (laughing) no/tu no/ voio mica/

The coherently isomorphic relation between Verb morphs and reference Pronoun system is also a characteristic of utterances where the child refers to himself. He may use either 3rd person Verb morphs and a Proper Name, or 1st person Verb morphs with 1st person Pronoun. Even when the variation is in the same discourse unit the relation is respected.

(3) Lele, 26 months of age (stage 8), pointing to a child who cries most of the time:
3.1. Lele: mato quelo là/
3.2. Int.: e perché sarebbe matto?
3.3. Lele: gioca mai/pange/Lele no pange/
3.4. Int.: qualche volta anche tu piangi/
3.5. Lele: pango mai io/

In the example (3) Lele switches from the use of his Proper Name (Lele) in 3.3. to personal Pronoun (io) in 3.5. to indicate himself: to the alternation Proper Name/personal Pronoun corresponds variation in the verbal flexive system (pange/pango).

In the same way, the 3rd person Verb morphs, singular and plural, alwyas refer to Subject Noun Phrases, singular or plural, or anaphoric Subject Pronouns.

(4) Elisabetta, 20 months of age (stage 3), pointing to a toy-horse in a corner:
4.1. Elisabetta: (l)o p(r)endo?
4.2. Int. : sì/ prendilo pure/
4.3. Elisabetta: (dragging the horse) eco/no voleva/

4.4. <u>Elisabetta</u>: venire/

(5) Orietta, 21 months of age (stage 4), pointing at a cat on the roof:
5.1. <u>Orietta</u>: gato nero/
5.2. <u>Int.</u> : che bel gattone! ti piacciono i gatti?
5.3. <u>Orietta</u>: (gr)afiano i gati?

From these examples it appears that almost all children's utterances are not only pragmatically appropriate in various ways (ERVIN TRIPP/MITCHELL KERNAN (eds) 1977) but also grammatically matched with respect to Verb morphology and case marking Pronouns. In other words, there is a one-to-one correspondence between the 1st person Verb morph and the 1st person Subject Pronoun, 2nd person Verb morph and 2nd person Subject Pronoun.

However, this regular use of Verb agreement with the Subject does not imply fixed order strategies. When the Subject is expressed, it may precede the Verb as in 3.3., or it may follow it as in 2.1.,3.5., 5.3. Table 5 specifies the curve of the SV order (including SV, SVO orders) and the VS one (including VS, VOS, VSO orders).

Table 5: percentages of Subjects preceding or following the Verb

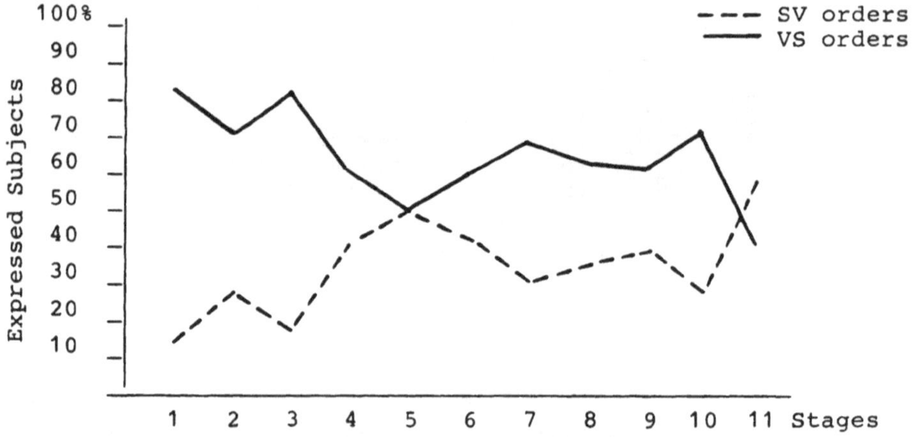

While guess and personal interpretation are essential for the identification of examples of GRs in some languages (McNEILL 1966 a: 119), in Italian and in Veneto dialect the formal system represented in the suffixal Verb conjugation and the pronominal

reference system allows the control of the mastering of GRs by "surface" means.

Despite the fact that the framework of intersecting morphosyntactic categories are rather different in Italian and in Veneto dialect (TEKAVČIĆ 1977),these coding properties are regularly and systematically mastered by the child from the very beginning of his linguistic production. However, this system of morphosyntactic rules that the child masters does not imply a fixed pattern in their linearization, which seems, according to Bates, only acquired as a whole by a later stage. Future work will concentrate on ordering rules that motivate variation encountered in different orderings. Such an explanation has to be given in terms of a theory of the text.

NOTES

1 CHOMSKY 1965.

2 See BOAS 1976.

3 From one year and three months to two years and nine months; from one year and four months to four years.

4 Contrary to what has been found elsewhere, the child does not necessarily use the 3rd singular person as the unmarked form, since the speaker and the hearer are always referred to by Proper Names instead of personal Pronouns (BATES 1976: 196).

5 In transcriptions of utterances I use the conventions generally adopted in conversational analysis both for adult and child discourse. The child utterances have been partially transcribed in phonetic notation: for instance, long consonants (geminates) pronounced as short in Veneto, have been transcribed as single segments. In the phonetic transcription I have utilized the graphic symbols of Italian in order to permit a more easily interpretation of the child utterance in relation with the preceding and/or following adult sentences.

REFERENCES

BARONI, Anna/FAVA, Elisabetta/TIRONDOLA, Gianna (1974): "L'ordine delle parole nel linguaggio infantile." MEDICI/SANGREGORIO (eds.) (1974): Fenomeni morfologici e sintattici nell'italiano contemporaneo. Atti del VI congresso internazionale di studi SLI. Roma 4-6 settembre 1972. Roma: Bulzoni: 375-390.

BATES, Elisabeth (1976):Language and Context.The Acquisition of Pragmatics. New York: Academic Press.

BOAS, Hans (1976) "Argumente gegen die Annahme linear geordneter syntaktischer Tiefenstrukturen in einer universellen Grammatik."BRAUNMUELLER/KUERSCHNER (eds.) (1975) Grammatik.Akten des 10. Linguistischen Kolloquiums. Tübingen 1975: Niemeyer: 129-138.

BOWERMAN, Melissa (1973a):Early Syntactic Development.A Cross Linguistic Study with Special Reference to Finnish. Cambridge: Cambridge University Press.

--- (1973b) "Structural Relations in Children Utterances." MOORE (ed.) (1973) Cognitive Development and the Acquisition of Language. New York: Academic Press: 197-214.

CHOMSKY, Noam (1965):Aspects of the Theory of Syntax. Cambridge: MIT Press.

COLE, Peter/SADOCK, Jerry (eds.) (1977):Syntax and Semantics.Grammatical Relations. New York: Academic Press.

ERVIN TRIPP, Susan/MITCHELL KERNAN, Claudia (eds.) (1977): Child Discourse. New York: Academic Press.

KEENAN, Edward (1976) "Towards a Universal Definition of 'Subject'." LI (ed.) (1976): 303-335.

LABOV, William (1972): Language in the Inner City. Philadelphia: University of Pennsylvania Press.

LI, Charles (ed.) (1976): Subject and Topic. New York: Academic Press.

Mc NEILL, David (1966a) "The Creation of Language by Children." LYONS/WALES (eds.) (1966): Psycholinguistic Papers. Edimburgh: Edimburgh University Press: 99-115.

--- (1966b) "Developmental Psycholinguistics." SMITH/MILLER (eds.) (1966): The Genesis of Language. Cambridge: MIT Press.

--- (1970): The Acquisition of Language. New York: Harper and Row.

MIONI, Alberto/TRUMPER, John (1977) "Per un'analisi del continuum linguistico veneto." SIMONE/RUGGIERO (eds.) (1977): Aspetti sociolinguistici dell'Italia contemporanea. Atti dell' VIII congresso SLI. Roma: Bulzoni: 323-372.

---/ARNUZZO LANSWERT, Anna (1977) "The Sociolinguistic Situation of Italy." Manuscript.

ROEPER, Thomas (1973a) "Theoretical Implications of Word Order, Topicalization and Inflection in German Language Acquisition." FERGUSON/SLOBIN (eds.) (1973): Studies in Child Language Development. New York: Holt, Rinehart and Winston: 541-553.

--- (1973b) "Connecting Children's Language and Linguistic Theory." MOORE (ed.) (1973): Cognitive Develompent and the Acquisition of Language. New York: Academic Press: 187-197.

SLOBIN, Dan (1966) "The Acquisition of Russian as a Native Language." SMITH/MILLER (eds.) (1966):Genesis of Language. Cambridge: MIT Press: 126-152.

TEKAVČIĆ, Pavao (1977) "L'indigeno e l'alloglotto nell'ordine delle parole di un idioma di frontiera: l'istroromanzo." CONTE/GIACALONE RAMAT/RAMAT (eds.) (1978): Akten des 12. Linguistischen Kolloquiums. Pavia 1977.

NATURAL TEXT PROCESSING AND WORD ORDER

Jan Wirrer

0. As far as research in natural text processing[1] is concerned word order belongs to the minor problems to be tackled. Thus - as far as I know - word order has hardly become an explicit topic in text processing research. Yet the problem is worthwhile to deal with in a short paper like this.

In my paper I will mainly discuss the following questions:

- What types of operations are included in the notion of natural text processing?
- How far is word order relevant to natural text processing?
- To what groups of persons, what kinds of texts or discourses, and what kinds of situations do my hypotheses about natural text processing and word order apply?

Due to the shortness of this paper and - partly - due to the research done so far none of these questions can be fully answered here. All I can do and all I want to do is to give a short survey and to make some suggestions.

1. The notion of natural text processing includes all operations by the means of which a natural speaker assigns a resultant or output text - for instance a text produced by a speaker on a book he has read - to an initiant or input text - for instance the book the speaker has read. All kinds of discourse can serve as initiant texts and resultant texts: newspaper articles, novels, review articles, poems etc. According to Wienold and Rieser (1978) and Petöfi (1978) I assume six text processing operations: referentialisation, paraphrasing, condensing, metatextual description, evaluation and explanation. For the purposes of this paper two operations are of particular interest: paraphrasing and condensation. By the notion of "paraphrasing" I understand all operations by whose means a natural speaker produces a resultant text

with the same intensional meaning as the input text so that no intensional informations are lost. That means, for example, that the same inferences can be drawn from both the resultant and the initiant text. This holds, e.g., for an initiant text like "Mary has got a tom-cat" and an output text like "Mary has got a male cat". As opposed to paraphrasing "condensing" includes all operations by the means of which a speaker reduces the intensional information of an initiant text so that some information is lost. Thus, a resultant text like "Mary has got a cat" is a condensate of an initiant text like "Mary has got a tom-cat".

2. Research in human memory should be taken into account by anyone who is interested in natural text-processing because memory is involved in any text processing operation. Although there exist a lot of very different models of human memory most psychologists who propose such models make a distinction between long-term memory and short-term memory or long-term store and short-term store respectively. This difference is not only a matter of duration in the sense that information entering short-term memory is deleted after a short period, whereas it is retained in long-term memory for a longer period. It is also based on the kind of linguistic information that enters short-term memory on the one hand and long-term memory on the other. Thus psychologists working in the field of text processing agree on the assumption that - in the normal state of affairs - it is semantic information or propositional content that enters long-term memory whereas syntactic, as well as phonological, and phonetic information only enters short-term memory and tends to be deleted after a period of 20 seconds or so. Thus Sachs (1967) and Jarvella (1971) have shown that syntactic structures of input paragraphs deteriorate with delay whereas propositional content is still retrievable. This does not fully hold for two experiments reported by Anderson (1974). Anderson shows that verbatim representation of inputs is available not only in immediate recall, but also - yet at a lower rate - in delayed recall tests. He concludes that immediately, the verbatim representation "is highly available producing good memory for exact wording. None-

theless, the propositional representation is sufficiently available to produce the propositional effects. At delays, however, the propositional representation comes to dominate and effects of sentence form diminish. This hypothesis comes close to obliterating the short-term vs. long-term memory distinction in favour of a model in which there are faster decay rates for the perceptual than for the propositional information"[2].

In the above quotation Anderson states that propositional representation has an effect on text processing right from the very beginning. This assumption might be supported by findings by Kintsch and J.M. Keenan[3]. In an experiment with the number of words per input sentence controlled and the number of underlying propositions varying they have shown that reading time for sentences is a function of the number of underlying propositions and not a function of the number of words in the surface structure. Still, these findings should be met with some doubt, because in some respects Kintsch's propositional representations prove linguistically and logically inadequate[4].

One might argue that - in the normal state of affairs - it is not the whole propositional content of an input text which is available at delays but only - according to van Dijk - the so-called macro-structure. Recent experiments reported by van Dijk et al.(1975), van Dijk/Kintsch (1975), Kintsch (1976), and Meyer (1975)[5] seem to support this hypothesis. But as the concept of macro-structure still needs a lot of elaboration[6] and as it is of minor importance for the very subject of this paper this problem can be neglected here.

As regards word order we first must distinguish between short-term memory and immediate recall on the one hand and long-term memory and delayed recall on the other. If verbatim representation roughly corresponds to short-term memory, word order should cause no problem as far as immediate recall is concerned. That is why short-term store need no longer concern us here. Some crucial points, however, arise when we consider the relations between word order on the one side and propositional representation, long-term store and delayed recall on the other.

But before I go into details I should mention the linguistic notion of "text base" which has recently been introduced in cognitive psychology. Kintsch, J.M. Keenan and McKoon[7] have shown in two experiments that not only explicit statements are integrated into the text base but also statements that must be inferred during reading in order to allow for a coherent interpretation of an input paragraph. Thus after a 15 minutes' delay subjects are unable to distinguish between statements actually presented and statements inferred from the input paragraph. This does probably not only hold for reading time inferences but also - as might be shown by story recalls - at least for some inferences drawn after reading of an input discourse.

If the dominant role of propositional content for discourse processing proves correct this has a decisive influence upon the rôle of word order in this domain. Thus - in the normal state of affairs - word order should be relevant to text processing, if it indicates that certain propositions underlie an input discourse while others do not. The crucial point is that - from an empirical point of view - we often simply do not know whether this is the case or not.

Before I discuss some examples, I should say what is meant by the statement "that a proposition or set of propositions underlies a sentence or a text". To begin with we must distinguish between proposition and implication. With respect to natural discourse processing a concept of implication less strict than material implication should be advocated. This is why Kintsch suggests the following definition: "P implies Q iff given a semantic memory, which contains a set of inference rules, and given a proposition P, Q is derivable from P."[8]

Implication as defined above "depends upon whose semantic memory and whose inference rules one is talking about. If a person does not have the necessary information or the required inference rules to derive A from B, then for that person B does not imply A, even though it might for someone else"[9]. - The following German sentences may serve as an example.

(1) Maria heiratet Karl.

(2) Karl wird von Maria geheiratet.

The propositions, that is the predicate argument structures, underlying these sentences are identical, whereas the implications to be drawn might not be. Sentence (2), e.g., might imply for some speakers that Karl is a henpecked husband while sentence (1) does not. For these speakers sentences (1) and (2) are not paraphrases of one another.

Now let us consider two sentences which only differ with respect to word order.

(3) Der Mann kauft das Auto.

(4) Das Auto kauft der Mann.

If these sentences are paraphrases for a speaker there should be no difference between them in long-term processing. The same applies to sentences (3) and (4) and their corresponding passives if speakers interpret the passive sentences both as paraphrases of one another <u>and</u> of sentences (3) and (4) as well.

How can we find out whether speakers really do process sentences such as (3) and (4) differently or equally? Let sentences (3) and (4) serve as target sentences for two randomly chosen groups of subjects. Then, all recall protocolls in which the predicate and the two arguments are reproduced are regarded as correct, protocolls with the predicate or one of the arguments missing are considered false. Correct recalls show that in both groups sentences (3) and (4) as well as their corresponding passives occur as a reproduction of the respective target sentence. Taking this into account we may maintain the hypothesis that the two target sentences are processed equally if the following conditions are fulfilled:

- The error rates of both groups do not show a statistically significant difference,

- the different frequencies of the sentences used as a recall of the targets do not turn out significant either.

If, however, these differences prove significant, we should come to the conclusion contrary to the null-hypothesis.

Yet natural speakers - including subjects - seldom produce paraphrases in the strict sense of the word.The most frequent operation with respect to discourse recall is condensing. Thus the following sentences are probably recalled like sentences (3) and (4) if there is no further reference to the man's age and the colour of the car in the input paragraph:

(5) Der junge Mann kauft das blaue Auto.

(6) Das blaue Auto kauft der junge Mann.

We again could conclude from these findings that - for the subjects tested - sentences (5) and (6) are not interpreted differently as far as the word order is concerned. We could not, of course, conclude that this holds also for sentences (3) and (6), for instance. However, if we could find out that the information about the man's age and the colour of the car is no longer retrievable for the subjects, we might assume that sentence (6), e.g., is stored and represented in long-term memory just as sentence (3).

There is still a crucial point with recalls that should not be overlooked. Recall experiments like that outlined above could show us whether subjects assign the same or a different meaning to input sentences with a different word order, but they, of course, tell us nothing about the meaning subjects actually do assign to the sentences. - This experiment, for instance, could make evident that the target sentences are interpreted differently, yet it could not tell us in what respect.

Here the concept of text base becomes particularly relevant to word order. As I have mentioned above inferences drawn from an input discourse are integrated into the text base. This applies not only to reading time inferences but probably to other inferences as well. If this assumption proves correct, subjects should draw distinct inferences from target sentences which have been interpreted differently. I will give an example. Let there be two short input paragraphs that differ only with regard to their respective target sentences which in turn are only distinguishable by their word orders. For example:

(7) Alle geladenen Gäste waren erschienen. Denn heute war ein großer Tag: Maria wurde von Karl geheiratet. Die Feier begann bereits am späten Vormittag und endete nach Mitternacht.

(8) Alle geladenen Gäste waren erschienen, denn heute war ein großer Tag: Von Karl wurde Maria geheiratet. Die Feier begann bereits am späten Vormittag und endete nach Mitternacht.

After presentation of the paragraphs the following test question could be put to the subjects after a delay of about 15 minutes: Did Mary have any more suitors besides Karl? - If the subjects to which the second paragraph has been presented give replies different from those of the other group - e.g. more positive answers - and if this difference proved statistically significant we might conclude that subjects give distinct interpretations to the two target sentences; and we also know in what respect these interpretations do differ.

But there is still another influence of word order on recall and discourse processing. Experiments reported by Engelkamp (1976), Hornby (1972), Bell (1976), and Anderson (1974) show that word order has an effect on processing which seems merely due to position. With respect to recall I will give the following example: If we scored the false recalls of the first experiments outlined in this paper we should probably find out that the noun phrase placed in the first position is reproduced more frequently than the noun-phrase in the second position irrespective of the position of the agent and the object. This might be due to the fact that speakers when they are presented active as well as passive sentences tend to interpret the concept placed in the first position as topic.

Finally results from a completion test reported by Bell make obvious that children aged between three and seven tend to produce active sentences whenever possible, even if this involves the addition of more words[10]. According to Bell these results are due to the so-called transformational complexity. An additional and an even better explanation for these findings might be the hypothesis that children prefer to produce sentences with the agent presented by the first noun-phrase even more than adults do.

As for processing of active and passive sentences two experiments reported by Anderson make obvious that it is more difficult to process passive than to process active sentences. Andersons's subjects were presented paragraphs with active and passive target sentences. After a two minutes' delay probes testing the target sentences were presented to the subjects. These probes were either actives or passives. The results of these experiments show that passive probes were verified more slowly than active probes. This finding might at least partly be due to word order and the subjects' inclination to expect the agent being represented by the noun phrase in the first position.

3. The phrase "in the normal state of affairs" which I have used several times should be interpreted in this very context with respect to texts, persons, and situations.

As for texts two aspects have to be taken into account. First: all that has been said above is only valid for prose not for poetry. This is due to special surface structure phenomena typical for poetry such as meter and rhyme. These phenomena might influence recall rather drastically. Second: strictly speaking the above hypotheses merely apply to written not to spoken language. This is why phenomena such as stress have not been taken into consideration.

The above assumptions about the influence of word order upon discourse processing probably only hold for adults and for children older than seven years or so. This should hold true if we take into account Bell's experiments which I have mentioned above.

Finally the above hypotheses only apply to so-called normal situations. Here all situations are excluded where learning by heart and literal quotation is required, because for these situations the distinction between long-term and short-term store as defined above can not be valid and should be discarded in favour of an alternative model of human memory defined in terms of levels of processing[11].

4. The hypotheses about the influence of word order on text processing developped throughout this paper have not yet been empirically tested. But I hope to make up for this in the near future[12]. - Nevertheless I hope to have shown how linguistic assumptions about word order can be based on empirical grounds.

ACKNOWLEDGEMENTS

1 For the distinction between natural text processing and computational text processing see Wirrer (1976).

2 Anderson (1974) p. 161. See also Craik/Lockhart (1972), who make evident that a more powerful framework of human memory defined in terms of levels of processing should be preferred to traditional multistore memory models. Yet for the very purposes of this paper a multistore model proves sufficiently powerful and offers even simpler explanations for the findings referred to in the following passages.

3 cf. Kintsch (1974).

4 cf. van Dijk (1977).

5 For a review article of this book see Mannhaupt/Wirrer (1977).

6 This is shown by Metzing (1977).

7 cf. Kintsch (1974).

8 Kintsch (1974) p. 57.

9 ibd.

10 cf. Bell (1976) p. 297.

11 cf. Craik/Lockhart (1972).

12 cf. Mannhaupt/Wirrer (in preparation).

REFERENCES

ANDERSSON, G.R. (1974) :"Verbatim and Propositional Representation in Immediate and Long-Term Memory". Journal of Verbal Learning and Verbal Behavior 13, 149.162.

BELL, Ph. (1976) :"Transformational Complexity and Sentence Difficulty: A Developmental Study ". Language and Speech 19, 285-302.

BURGHART, W. / HÖLKER, K. (eds.) : Text Processing. Papers on Text Analysis and Text Description. (to appear in 1978).

CRAIK, F.I.M. / LOCKHART, R.S. (1972) :"Levels of Processing: A Framework for Memory Research". Journal of Verbal Learning and Verbal Behavior 11, 671-684.

van DIJK, T.A. et al. (1975) : Recalling and Summarizing Complex Discourse. Mimeo. University of Amsterdam.

van DIJK, T.A. (1977) :"Diskursbedeutung und Gedächtnis .Rezension von W. Kintsch, The Representation of Meaning in Memory".Wirrer, J. (ed.) 1-49.

ENGELKAMP, J. (1976) : Satz und Bedeutung. Stuttgart.

HORNBY, P.A. (1972) :"The Psychological Subject and Predicate" Cognitive Psychology 3, 632-642.

JARVELLA, R.J. (1971) :"Syntactic Processing in Connected Speech". Journal of Verbal Learning and Verbal Behavior 10, 409-416.

KINTSCH, W. (1974) : The Representation of Meaning in Memory. Hillsdale.

KINTSCH, W. / van DIJK, T.A. (1975) :"Comment on se rappelle et on résume des histoires ". Langage 40, 98-116.

KINTSCH, W. (1976) :"Comprehension and Memory of Text". (To be published in: Estes, W.K. (ed.), Handbook of Learning and Cognitive Processes, vol. 5. Hillsdale.)

MANNHAUPT, H.R. / WIRRER, J. (1977) :"Semantische Textstrukturen und Empirie". Bemerkungen zu B.F. Meyers Buch "The Organization of Prose and its Effect on Memory". Wirrer, J. (ed.), 50-108.

MANNHAUPT, H.R. / WIRRER, J. (in preparation) : Zum Einfluß der Wortstellung auf die semantische Verarbeitung.

METZING, D. (1977) :"Makrostrukturen als Element von Teun A. van Dijks Textgrammatikprogramm". Wirrer, J. (ed.), 109-128.

PETÖFI, J.S. (1978) :"Bemerkung zu Wienold, G. / Rieser, H., Vorüberlegungen zur Rolle des Konzepts der Textverarbeitung beim Aufbau einer empirischen Sprachtheorie". Burghart, W. / Hölker, K. (eds.).

SACHS, J. (1967) :"Recognition Memory of Syntactic and Semantic Aspects of Connected Discourse". Perception and Psychophysics 2, 437-442.

WIENOLD, G. / RIESER, H. (1978) :"Vorüberlegungen zur Rolle des Konzepts der Textverarbeitung beim Aufbau einer empirischen Sprachtheorie". Burghart, W. / Hölker, K. (eds.)

WIRRER, J. (1976) : Kleines Lexikon der Linguistik. Stichwort: Natürliche Textverarbeitung. Linguistik und Didaktik 27, 247-250.

WIRRER, J. (ed.) (1977) : Textgrammatische Konzepte und Empirie. Hamburg.

L'ORDINE DELLE PAROLE COME ELEMENTO DI TRADIZIONE E DI MEMORIZZAZIONE

Giulia Petracco Sicardi

0.1. Nelle lingue flessive con sistema morfosintattico fortemente articolato l'ordine delle parole non è pertinente per l'espressione delle opposizioni e dei rapporti morfosintattici, essendo per questa funzione sufficienti i morfemi. In alcune lingue di questo tipo tuttavia si riscontra che, se l'ordine delle parole è libero, esistono però posizioni a b i t u a l i fisse di alcuni elementi della proposizione, per esempio il predicato alla fine dell'enunciato, l'attributo davanti al sostantivo, l'avverbio e l'oggetto davanti al verbo, il genitivo prima del sostantivo, la negazione davanti all'elemento negato. In queste condizioni l'ordine delle parole si configura come elemento di tradizione più che come tratto funzionale. Non importa qui stabilire l'origine di questi elementi di tradizione. Il problema che intendo proporre è quale sia l'utilizzazione f u n z i o n a l e dell'ordine tradizionale o abituale delle parole e quanto questa favorisca la conservazione dei modelli e quindi anche la stabilità dell'ordine delle parole.

Si è finora data importanza soprattutto alla funzione e s p r e s s i v a dell'ordine delle parole, quando esso non è grammaticalmente pertinente. Cercherò invece di mettere in luce la funzione di memorizzazione dell'ordine fisso. Non ritengo che tale funzione sia primaria, bensì che subentri secondariamente quando il testo è di tradizione orale o in situazioni affini, ad esempio quando l'apprendimento di formulari avviene nella scuola attraverso la m e m o r i z z a z i o n e[1]. Ma quando la funzione di memorizzazione è in atto, si stabilisce una interazione tra le due funzioni: la frequenza crescente di un modello sostiene e rafforza il valore espressivo delle diverse posizioni nell'enunciato e conserva l'ordine abituale delle parole; d'altra parte la ripetizione di determinate clausole agisce come elemento di richiamo (analogamente a quanto avviene per la rima) e viene quindi utilizzato sempre di più, a scàpito di altri elementi concorrenti, come elemento fondamentale nel processo di memorizzazione.

0.2. Ho scelto come materiale da analizzare i proverbi lituani, perché costituiscono un corpus omogeneo di testi di tradizione orale e presentano un ordine delle parole particolarmente rigido. Dei circa 4000 proverbi che si conoscono[2] la piccola raccolta di 290 proverbi, che è inserita nel Lesebuch del Senn[3], non rappresenta un vero campione, perché non conosciamo i criteri della scelta. Quindi le mie osservazioni, che riflettono l'analisi della raccolta Senn, hanno soltanto carattere esemplificativo.

1.1. La maggior parte dei proverbi analizzati si può ricondurre a tre tipi principali: I. un enunciato quasi sempre articolato in tre membri; II. due enunciati a costruzione parallela; III. una locuzione comparativa.[4]

Il I tipo è caratterizzato da una struttura bipolare. La posizione iniziale è riservata all'elemento che definisce l'argomento (quasi sempre il soggetto = S). La posizione finale è riservata al predicato (= P, che spesso è un verbo, ma può anche essere un elemento nominale, sostantivo o aggettivo o participio o infinito). Indicando con O l'oggetto diretto, con I l'oggetto indiretto o altro determinativo del verbo, abbiamo il modello SOP, come:

(57) ãklas ãklą vedžiója 'il cieco conduce il cieco'

e il modello SIP, come:

(18) ir gaidỹs sàvo kiemè drąsùs 'anche un gallo è coraggioso in casa sua'

(62) áuksas ir purvè žìba 'l'oro brilla anche nel fango'.

1.2. Se O o I costituiscono il tema del discorso, prendono il posto di S:

(58) alkanám dúona rūpi 'all'affamato il pane dà preoccupazione'

(123) liežùvio saũso / niẽkas / neklaũso 'la lingua arida nessuno l'ascolta'.

Assimilabili a queste strutture di ritmo ternario sono i proverbi, in cui il soggetto non è espresso, perché il verbo ha valore impersonale (e in questa funzione può essere espresso nella III o nella II persona singolare). Es.:

(88) iš ãdatos vežimą priskáldo 'con un ago il carro si spacca'

(165) ant svẽtimo árklio / netolì / tejósi 'su un cavallo straniero non cavalcherai lontano'.

2.1. Il II tipo ha una struttura binaria con due membri paralleli. Il modello è SOP/SOP, SP/SP, IP/IP ecc.:

(59) ankstyvàsis dantìs rakinė̃ja / vėlyvàsis akìs krapštinė̃ja 'il dente del veloce stuzzica, l'occhio del lento raschia' (SP/SP)

(64) bėdà bė̃dą vẽja / var̃gas var̃gą rẽmia 'il bisogno scaccia il bisogno, la miseria rafforza la miseria' (SOP/SOP)

(170) be tiñklo nesugáusi / be giñklo nenušáusi 'non puoi acchiappare senza rete, non puoi uccidere senz'arma' (IP/IP).

Ma spesso un elemento è comune ai due membri e quindi non viene ripetuto, come:

(71) didžių̃ vagių̃ niẽkas nekùšina / mažúosius kársto 'i ladri grossi nessuno li tocca, i piccoli li appendono' (OSP/OP)

(113) kiaũlė laukè dvõkia / stalè kvẽpia 'il maiale nel campo puzza, sulla tavola olezza' (SIP/IP).

In tutti questi esempi i due membri sono uniti per asindeto.

2.2. Altri proverbi del II tipo hanno i due membri collegati da pronomi, avverbi o locuzioni correlative, come:

(97) k a i p móku / t a i p šóku 'come insegni, così salti'

(155) v i e n ì dýgsta / k i t ì ñyksta 'gli uni germogliano, gli altri spariscono' (SP/SP).

Altri proverbi hanno i due membri uniti dal pronome relativo (eventualmente richiamato nel secondo membro dal pronome dimostrativo o anaforico), legame che è modellato sulla correlazione, come:

(103) k a s daũg žãda / t a s mãža dúoda 'chi promette molto, fa poco' (SIP/SIP)

(190) k a s ar̃s / nepavar̃gs 'chi arerà, non diventerà povero' (SP/P).

Una menzione particolare meritano i proverbi a due membri che hanno il primo membro con un verbo di modo indefinito, in genere un participio. Resta qualche incertezza se questo gruppo vada analizzato come ritmo binario o ternario, potendo il participio congiunto costituire un'unità fonica con il sostantivo. Ritengo però che sia più soddisfacente l'analisi binaria, dato che il participio occupa per lo più il secondo posto nel primo membro, sul modello SP:

(164) ãklas bū́damas / kitíems kẽlio neparódysi 'essendo cieco, non può mostrare ad altri la strada' (SP/IOP)

(168) ãvinę d ú o t a s / dar gaĩdžio geĩdžia 'chi ha avuto in dono un becco, desidera anche un gallo' (OP/OP).

Abbiamo infine un gruppo a struttura binaria con due participi, come:

(222) geriaũ n e p r a d ė́ j u s / nekaĩp n e p a b a ĩ g u s 'meglio chi non ha cominciato che chi non ha finito'.

3.1. Lo schema del III tipo è costituito da un aggettivo o participio oppure da un verbo, seguito da <u>kaip</u> 'come' e da una locuzione o proposizione comparativa (= C); cfr. it. <u>nero come la pece</u>; <u>corre come una lepre</u>. Non si tratta quindi di proverbi veri e propri, bensì di similitudini ellittiche, riferite ad un attributo (= A) o a un verbo (= V). L'ordine delle parole è rigido: A/kaip/C o V/kaip/C. Esempi:

(4) ãtbulas kaip vėžỹs 'che cammina all'indietro come un gambero'

(82) gìrgžda kaip rãtai neteptì 'scricchiola come ruote non unte'

(121) láukia kaip gérvė giedrõs 'aspetta come la gru il bel tempo'.

Su 290 proverbi analizzati abbiamo 28 esempi del III tipo. Poiché questo modello ritorna in molte lingue, non fornisce elementi particolarmente perspicui per l'ordine delle parole, salvo confermare la struttura bipolare con preminenza della posizione iniziale rispetto alla posizione finale.

4.1. Nei modelli descritti fin qui il predicato occupa la posizione finale.[5] Questa posizione è nettamente prevalente, anche se le eccezioni alla norma sono abbastanza numerose: 27 contro 85 nei proverbi del I tipo e 46 contro 98 in quelli del II tipo.[6]

Il verbo all'imperativo occupa normalmente la posizione iniziale, come:

(262) a t e ĩ k beláukiamas / i š e ĩ k bemýlimas! 'arriva ancora atteso, parti ancora amato!'

(263) b ū́ k gìrtinas / bet n e s i g ì r k ! 'sii amato, ma non amarti!'

(279) n e g ì r k dienõs be vãkaro! 'non lodare il giorno prima di sera!'

Ma se l'imperativo è nel II membro, il verbo va in posizione finale, come:

(274) kas táu negẽra / ir kitám n e d a r ý k ! 'ciò che non è buono per te, non farlo agli altri!'

Da segnalare un gruppo di proverbi (ad es. 264, 267, 270, 280-3) che presentano l'imperativo nel I membro e nel II membro hanno un futuro che esprime la conseguenza prevista. Mentre l'imperativo occupa la posizione iniziale, la posizione del secondo verbo è libera:

(264) d ė́ k grū́dą prie grū́do / p r i p ì l s i aruoda 'poni grano dopo grano, riempirai una cassa'

(278) n e b ū́ k per drasùs / akìs i š d è g s i // n e b ū́ k per bẽtas / daliẽs n e t è k s i 'non essere troppo coraggioso, l'occhio si spegnerà; non essere troppo prudente, perderai la parte'.[7]

Evidentemente anche il proverbio a struttura parallela è sentito come una unità; perciò l'imperativo occupa la posizione iniziale o quella finale (sempre la posizione di maggior rilievo) a seconda che si trovi nel I o nel II membro.

4.2. Oltre all'imperativo, si possono rilevare altre sei condizioni che determinano deviazioni dalla norma della posizione finale del predicato.

In 56, 60, 93, 107, 112, 184, 225, 233, anziché il predicato, si trova in posizione finale un elemento che presenta rima o assonanza con l'elemento finale del I membro. Ad es.:

(56) žìlas pláukas g a l v o j è / vélnias u o d e g o j è 'capelli grigi sulla testa, il diavolo nel codazzo'

(184) iš tų̃ š i a u d ų̃ nebùs g r ū d ų̃ 'dalle festuche non verrà grano'.

In 5, 6, 38, 42, 46, 49, 66, 74, 98, 158, 187, 196, 219, 250, 270 sta in posizione finale un infinito o un participio o un nome verbale, anziché il verbo coniugato che in qualche caso è un verbo copulativo o servile. Es.:

(5) bèdà su kvailù b á r t i s / su plikù p è š t i s '(è) difficile litigare con uno stupido, accapigliarsi con un calvo'

(74) dúona ver̃kia tìnginio v á l g o m a 'il pane piange (se è) mangiato dal fannullone'

(187) kad daũg norési ž i n ó t / greĩt pasénsi 'se vorrai conoscere molto, invecchierai presto'.

In 55, 80, 217, 245 il predicato è posto in posizione iniziale, talora con funzione espressiva. Es.:

(217) atitìko kir̃vis kótą 'va d'accordo l'ascia con il manico'.

In 32, 33, 34, 35, 36 il predicato nominale si trova in posizione iniziale, perché è un elemento correlativo. Es.:

(33) kokià obelìs / tokiẽ ir obuoliaĩ 'quale il melo, tali anche le mele'.

Se il predicato è un elemento nominale (come in 6, 9, 10, 17, 21, 27, 41, 45, 49, 52, 54, 55, 66, 90), è facile che S e P si scambino la posizione. Es.:

(10) devynì amataĩ / dešim̃tas bãdas 'nove i mestieri, il decimo (è) la fame'

(17) ìlgas pláukas / trum̃pas prõtas 'lunga capigliatura, corta mente'

(55) žilà kiaũlė / žilì ir paršẽliai 'grigio il maiale, grigi anche i porcelli'.

In 14 e 50 l'aggettivo comparativo occupa la posizione iniziale. Questo caso è assimilabile ai costrutti comparativi del III tipo, che hanno appunto P all'inizio:

(14) gerèsnis žvìrblis rañkoje / nekaĩp élnias gìrioje '(è) migliore il passero nella mano che il cervo nella foresta'

(50) sunkèsnis vežìmas priešˇ kálną '(è) più difficile il trasporto verso la montagna'.

I proverbi con il verbo in posizione non finale che non rientrano in questa casistica sono pochi (7, 8, 51, 87, 95, 98, 111, 118, 133, 151, 221, 266) e potrebbero, almeno in parte, riflettere una tradizione diversa, minoritaria rispetto ai modelli e alle norme che ho esposto.

5.1. Dall'analisi condotta derivano alcune considerazioni di carattere generale. L'ordine abituale SP è rigido. L'elemento nominale del predicato e le forme nominali del verbo sono trattate generalmente come il verbo. Il fatto è significativo, data l'importanza che hanno le costruzioni nominali nei proverbi lituani.

La posizione iniziale è di maggior risalto rispetto alla posizione finale. Quindi in posizione iniziale va l'elemento che è considerato più significativo. Invece la posizione finale del predicato non è dovuta alla scelta bensì alla tradizione: è una posizione abituale come quella dell'attributo davanti al sostantivo.

La rigidità rende l'ordine delle parole un elemento di memorizzazione cioè di richiamo. Esso non è però l'unico. Abbiamo anche gli elementi correlativi, la rima, l'allitterazione. Questi quattro elementi talora concorrono, ma per lo più sono alternativi. Si può stabilire questa graduatoria di importanza degli elementi di richiamo: 1. elementi correlativi, 2. ordine delle parole, 3. rima, 4. allitterazione. Concordano in posizione iniziale S, elementi correlativi e allitterazione; in posizione finale, P e rima. La formazione dei proverbi avviene per riproduzione di modelli, per sostituzione o per aggiunta di elementi nello stesso modello.

5.2. Sull'ordine delle parole in lituano le opinioni non sono concordi. Kurschat[8] ne ha dato una definizione rigidissima: soggetto e attributi, dativo dell'oggetto indiretto, avverbio, accusativo, complementi prepositivi del verbo, verbo. L'analisi di Kurschat è limitata alle proposizioni indipendenti. Schwentner[9], che estende l'analisi alle proposizioni dipendenti e alle frasi interrogative, sostiene che sono presenti le tre posizioni del verbo: finale, iniziale e mediana. Schwentner però non distingue il tipo di testo che cita, limitandosi ad escludere le traduzioni[10].

Ritengo che l'ordine delle parole vada studiato analizzando materiali omogenei. Questo sempre, e in particolare per il lituano, la cui produzione scritta è molto diversificata. L'analisi condotta sul corpus dei proverbi mi pare che dimostri l'esistenza di modelli con alto indice di frequenza. Qualche sondaggio fatto sulle dainos (liriche popolari) e sulle raudos (canti popolari) hanno rivelato modelli diversi: le raudos risultano più affini ai proverbi che le dainos.

NOTE

1 Questo è il caso, ad esempio, del latino notarile altomedioevale, che è caratterizzato dalla ripetizione meccanica di formule mandate a memoria. Per ragioni di spazio rimando ad altra occasione l'analisi di questo tipo di testi, che si presenta interessante per lo studio dei modelli sintattici nel passaggio dal latino classico al latino medioevale.

2 Pubblicati a varie riprese, cfr. SENN 1957, p.15.

3 SENN 1957, pp.15-22. Negli esempi citerò i proverbi con il numero d'ordine che hanno in questa raccolta.

4 Sulla sintassi dei proverbi lituani è in corso di stampa un mio contributo in "Studi slavistici in ricordo di Carlo Verdiani" a cura di A.M. Raffo, Pisa, Giardini.

5 In alcuni casi (ad es. 24, 27, 37, 42, 46, 52) funge da predicato un complemento indiretto o un avverbio. In altri casi (ad es. 15, 20, 21, 31, 46, 268) il predicato non è espresso.

6 Ho considerato nelle eccezioni tutti i proverbi del II tipo, in cui uno dei due verbi non fosse in posizione finale. Nelle locuzioni del III tipo il predicato (nominale o verbale), che è anche il referente, è sempre in posizione iniziale.

7 Come si vede da questo esempio, anche i pochi proverbi (109, 177, 191, 230, 238, 265) formati da più di due proposizioni confermano l'importanza delle norme rilevate, quella della struttura binaria o ternaria del proverbio e quella della posizione finale del predicato.

8 "Grammatik der litauischen Sprache", 1876, pp.440 sgg., cit. da SCHWENTNER 1922, p. 9.

9 SCHWENTNER 1922.

10 La fonte delle citazioni è quasi sempre l'antologia di testi contenuta in Wiedemann, "Handbuch der litauischen Sprache", Strasburgo, 1897.

BIBLIOGRAFIA

SCHWENTNER, Ernst (1922): Die Wortfolge im Litauischen. Heidelberg, Winter.

SENN, Alfred (1957): Handbuch der litauischen Sprache. II.Lesebuch u. Glossar. Heidelberg, Winter.

--- (1966): Handbuch der litauischen Sprache. I. Grammatik. Heidelberg, Winter.

SPANISH WORD ORDER. SUBJECT POSITION. DISCUSSION OF SOME FORMAL CONSTRAINTS

Nicole Delbecque

In this paper I will not go into the traditional accounts of Spanish word order. Let me just recall that most grammars give intuitive explanations for the generally observed variation in the subject position. Clarity, emphasis, the topic-comment distinction, the stylistic notion of hyperbaton and the 'general freedom' in the Spanish language are the key notions which underlie most descriptions of the order of constituents.

I do not deny that semantic differences and stylistic or pragmatic reasons may motivate to a certain extent the distinct positions. In fact, the morpho-syntactic structure of the sentence as well as the thematic-functional organization of the utterance have to be taken into account in order to understand fully the problem of the collocation of the major constituents. The question whether the description has to be based more heavily on semantic features, rather than on formal features, remains unclear. In the litterature one finds a number of very valuable contributions which mainly stress the semantic aspects of the problem concerned. In order to find out which is the most appropriate way to incorporate this aspect of the Spanish language into its grammar, a thorough investigation of all the possible morpho-syntactic features is needed.

The basic questions I am concerned with are the following. What does it mean to say that the order subject-verb-object is the dominant order for Spanish ? And, is it possible to invoke other than semantic, stylistic or pragmatic reasons in order to account for the generally observed variation ? My aim is to demonstrate that the so-called freedom in the ordering of the sentence constituents obeys to some formal constraints. In other words, without underestimating the explanatory power of semantic and pragmatic analyses, I want to stress the importance of morpho-syntactic features in this domain.

First let me present an overall picture of the facts related to the position of the verb and its grammatical subject. Momentarily I exclude the sentences without overt grammatical subject; I will comment them briefly at the end. As far as the verb position is concerned, it can be said that in half the cases the verb opens the sentence, and in one case out of twenty the verb comes in final position. These proportions are based on a close examination of two modern essay writers : Ortega y Gasset and Unamuno[1].

This gross classification of the verb position thus accounts for 65 % of the material. For now this may suffice, in order to relate the verb position to the subject position. Of course, initial verb position triggers automatically the postposition of the subject, and final verb position the anteposition of the subject. This strict correlation holds for 15 % of the material (10 % for subject postposition, 5 % for subject anteposition). Nevertheless, the subject position cannot be described in terms of the verb position, since in up to 26 % of the cases (297 of the 910) in which the verb occupies the second position the subject is postposed. The verb second position thus coincides for no more than 74 % of the cases with subject anteposition, i.e. in 7 cases out of 10. (When a clitic forms part of the verbal syntagm the proportion anteposed vs. postposed is a bit lower than 7 to 3).

In our data, 7 out of 10 appears to be the overall proportion of subject anteposition. It now becomes clear to what it really amounts to say that the subject-verb order is the dominant order : in prose writings 70 % of the sentences present this order. And in eight cases out of ten with anteposed subject the sequence subject-verb opens the sentence. To these initial data a final typological feature can be added concerning the postposed subject : in eight cases out of ten it stands in final position. We know thus the relative frequencies of both anteposition and postposition, and we can construct the skeleton of the type-sentence with an internal position for the verb and an external position for the subject.

Now the analysis can begin. It has to be made explicit which type of syntactic structures allow for variation in the position of the subject, and, more precisely, which features are connected with subject postposition rather than anteposition, and vice versa. My hypothesis is that a number of factors, formal factors, are related to the subject position; in other words, that the so-called 'free variation' is conditioned by systematic features. The aim of the investigation is to develop the 'variable rule' for the position of the grammatical subject vis-à-vis its verb. Up till now the variation in the position of the subject has been viewed as a kind of free variation, not bound by any kind of formal, intralinguistic features. Instead of relying upon factors as vague and ambiguous as expressivity, variety, style, and such more, I suggest that the position of the subject is for a great deal the result of interwoven linguistic features. I presume that a whole lot of linguistic constraints can affect in different degrees the subjectposition. Some characteristics will favor postposition, others will disfavor it.

As I have already mentioned, I concentrate uniquely on a few formal features. There is, however, no doubt in my mind as to the relevance of the topic-comment articulation and related pragmatic matters for gaining more insight in the sequencing of the syntagms. Heles Contreras discusses the problem extensively from this point of view in his study "A theory of word order with special reference to Spanish", published last year[2].

By the very fact that I am applying a quantitative method, which is automatizable, I feel obliged to say a few words about the procedure followed. I will exemplify it referring to the essay collection of Ortega y Gasset, "Espíritu de la Letra". This corpus is made up of 2783 sentences with a finite verb. Grammatically independent sequences without overt finite verb and quotations are not considered. These sentences, which constitute the corpus, are all subject to the same analysis. It is crucial that the analysis be exhaustive. The analyzed aspects, call them variables, receive a consistent codification, ready to be punched and to form the input for later computer operations. This first step is purely analytic. In synthesizing I have distinguished three steps. First, it is necessary to get a more or less precise idea about the occurrence of the distinct variables. Here I rely upon the absolute and relative frequencies, automatically computed. Second, the interrelation between the factor groups, the groups of variables, has to be measured; for instance, how many times is it the case that a four-constituent sentence containing a transitive non-pronominal verb and lacking any introductory syntagm before the subject-verb unit, has its subject postposed ... how many times has it the subject anteposed ... and, is this proportion deviant from the overall one ? This kind of questions can be answered by computing cross-tabulations, similar to those used in social sciences, in function of the variable under consideration, in this case : the subject position. In comparing cross-tabulations of various combinations of variables, relevant tendencies show up. In order to unify the observations, and, thus, to find a way of accounting for the observed tendencies in a connex manner, a last program can be run. Herefore I use the Cedergren-Sankoff computer program, put at first into use for socio-linguistic research. This program enables me to weigh the chosen factors with respect to the observed variation. It was especially designed for this kind of linguistic research where the aim is to account for systematic variation and to describe which factors, and in which measure, determine the selection of one variant rather than another. This program is based on probability theory. It estimates for every factor the maximum likelihood of application of the variant, which in this case is the postposition of the subject. Hence conclusions can be drawn which go well beyond the frequencies observed in the actual data to predictions on a much larger scale. As Labov and others have pointed out, this way of viewing linguistic variation has implications for linguistic theory. It has turned out to be necessary to define a new kind of grammatical rule, which is as much part of the language structure and of the individual competence as the former categorical and optional rules : the variable rule.

The results I will present now will thus lead to a tentative formulation of the variable rule for the subject position in Spanish sentences.
In order to have a comparative basis for my results, I have preferred to discuss those features for which I had already made a similar, though less refined, analysis earlier on another contemporaneous corpus.

As a first constraint I would like to present the sentence length, measured by the number of sentence constituents. This feature had already been studied in a poem collection of Unamuno, "Cancionero"[3]; this is a corpus of 8713 sentences, 3574 of which have an overt subject. There it is the case that the probability of encountering a postposed subject is the highest in the shortest sentences, and the lowest in the longest sentences, with a steady degression. In the essay collection of Ortega y Gasset, the same tendency shows up. The figures in table 1 represent it. The appearance of probability 1 for sentences with two constituents in Unamuno does not mean that in these cases the postposition occurs in 100 % of the cases; the first version of the program was so designed as to assign the figure 1 to the most relevant factor for the variation concerned. The modified version I am using now, just assigns a probability between zero and one to the different factors, and seems to fit the data better. Thus the more sentence constituents, the lower the probability of having the subject postposed. Note, however, that for sentences with three and four constituents the probability is the same in Ortega y Gasset. This does not detract anything from the tendency in its totality.

Number of constituents	2	3	4	5	6+
Ortega y Gasset (prose)	.88	.59	.37	.37	.34
Unamuno (poetry)	1. (.53)	.91 (.51)	.83 (.48)	.82 (.48)	/

Table 1. Effect of the number of constituents on the postposition of the subject in "Espiritu de la Letra", essay collection of Ortega y Gasset and "Cancionero", poem collection of Unamuno.

Remarkable is the fact that this tendency can be observed in poetry as well as in prose. The fact that the probabilities for Unamuno's "Cancionero" are generally higher, close to one, is due to the different relative frequency of postposition and anteposition. Later on I will comment briefly upon this difference, while comparing both corpora in the light of the global results.

In october '77 the second version of the program has also been applied to the formerly analyzed Unamuno-corpus in order to provide a wholly consistent comparison between the results of both corpora. These new figures have been added between parentheses in all tables. They do not require any specific comment; therefore they merely serve as an illustration of the different design of both versions of the Cedergren-Sankoff program.

The second feature I want to discuss concerns the type of verb. By this I do not refer to semantic categories, but to syntactic classes. For Spanish six classes can be distinguished : the non-pronominal transitive verbs with an expressed complement (NP, PP or S) (tiene una casa : he possesses a house), the non-pronominal transitive verb without complement in surface (paga : he pays), the pronominal transitive verbs (se mata : he kills himself), the non-pronominal intransitive verbs (anda : he walks), the pronominal intransitive verbs (se muere : he dies), and the class constituted by SER and ESTAR, the copulas. In table 2 these classes come ordered according to the probability of having their grammatical subject postposed. The feature (+pro) appears to be the most salient one; the next important one is (-tr). By retaining the features (+pro) and (-tr) as the relevant ones I am interpreting the fact that verbs which present the features (+pro) present the highest probability, the verbs which are (+pro) and (-tr) the second highest, and the verbs which are (-pro) and (-tr) the third highest. The copulas sort of form the division line between the verbs of high probability (higher than .5) and those of low probability (lower than .5). Along the same lines it is interesting to see that within the verbs which are the least likely to have their subject postposed, the transitive verbs used 'intransitively' - in traditional wording - still present a higher probability than the 'real' transitives.

Type of verb	(+pro) (+tr)	(+pro) (-tr)	(-pro) (-tr)	copula	(-pro) (+tr) (-d.o.)	(-pro) (+tr) (+d.o.)
Ortega y Gasset (prose)	.64	.54	.51	.47	.42	.41
Unamuno (poetry)	(-tr) 1. (.57)		(+tr) .72 (.43)			

Table 2. Effect of the type of verb on the subject postposition. The following features are used to distinguish distinct verb classes : (pronominal) (transitive), (direct object) and (copula).

The relevance of the feature (+pro) has already appeared in an earlier study of Unamuno's newspaper articles where the subject of a pronominal verb follows the verb four times more than the subject of a non-pronominal verb[4].

As is shown on table 2, the importance of the feature (-tr) is confirmed in Unamuno's poetry. In this poetic corpus I have not yet checked the relevance of the feature (+pro). Anyway, it can be stated that the actual results all bring confirmation of earlier partial analyses. It thus seems plausible to hypothesize that, whichever the literary genre, these syntactic features (+pro)(-tr), will favor the postposition of the subject.

As a last feature I would like to discuss the correlation between the kind of introductory constituent and the subject position. By introductory constituent I mean the following classes, distinguished as much as possible on formal grounds, and partly on the basis of their grammatical function, too. Formally defined are the adverbial phrase, the prepositional phrase, the clitic pronoun ... also grammatically defined are the NP direct object, the PP direct object, the relative pronoun direct object, the sentential direct object, the NP predicative complement, and the NP attributive complement (to the subject or the object).

The presence of a predicative complement before the verb is most likely to co-occur with a postposed subject. The predicative complement may stand alone or be accompanied by any other syntagm (PP, AdvP). This high probability, of .73, thus applies to the copulas.

The second most relevant openings are constituted by a PP preceded by an AdvP. When a sentence begins with a PP, either preceded by an AdvP or followed by a clitic pronoun, the subject can be expected to be in postposition.

A direct object preceding the verb, be it NP or PP, alone or with other syntagms, triggers also the postposition of the subject, with a probability of .63. When the direct object appears as an entire sentence, the probability is still high : .61.

A sequence of at least two AdvP triggers postposition with a higher probability than a sequence of at least two PP's : .59 versus .55. And with a probability of .53, an AdvP followed by a clitic pronoun seems a little more relevant than just a PP, the last with a probability of .53.

The feature weight of the sequence containing a PP followed by an AdvP approaches the limit of what can be called significative; the probability of having the subject postposed is no higher than .52.

Ortega y Gasset (prose)	(X)-Pred-(Y) .73 PP-Clitic .66 AdvP-PP .65 (X)-PP/NP-(Y) .63 S .61	(X)-AdvP-AdvP-(Y) .59 (X)-PP-PP-(Y) .55 AdvP-Clitic .53 PP .53 PP-AdvP-(X) .52	Rel.Pro-(X) .41 AdvP .39 Attrib.-(X) .33 Clitic .25 Ø .18			
Unamuno (poetry)	NP	≥2	PP	Clitic	Ø	
	1.	.93	.92	.88	.86	
	(.54)	(.51)	(.49)	(.48)	(.47)	

Table 3. Effect of the introductors on the subject postpostion. (X) and (Y) indicate the facultative presence of other syntagms. Pred refers to the predicative complement. PP to the prepositional phrase, AdvP to the adverbial phrase, PP/NP to the prepositional phrase or the noun phrase which function as direct object, and S to the clause.

The remaining introductory sequences are more likely to co-occur with subject anteposition. The probability of having the subject anteposed is the highest when there is no introductory syntagm at all (the probability figure here is .18). When it is constituted uniquely by a clitic or a sequence of clitics the anteposition is very likely, too : the probability of postposition is as low as .25.

Any attributive complement to an NP (be it subject or object) is not relevant for subject position. This means that an NP introduced by como ('as') or either a gerund or a participle separated from its head, do not influence the subject position. The presence of only an AdvP before the verb is not to be reckoned with, either.

Whereas postposition appears to correlate highly with the anteposition of the direct object in the form of an NP, a PP or a S (clause), this is not the case for the relative clauses where the relative pronoun functions as direct object. The relative pronoun, whether or not followed by other syntagms, seems indifferent to subject position.

Some time ago a rough analysis of the introductors in Unamuno's "Cancionero" has given the results represented in table 3. Being less complete (by exclusion of the copulas, for instance), and less accurate (by collapsing all introductory sequences constituted by two or more constituents), those results can only be compared partially with the actual ones. For both materials the absence of introductors least favors postposition, and a clitic is assigned the second lowest probability of having the subject postposed. Looking at the probability of .93 for two or more introductors, it could be said that this falls in with the high probability of having a postposed subject when the introductory sequence is composed by a PP and a clitic, an Avdp an a PP, and the like.

The fact that in the previous analysis the presence of an NP direct object has presented the highest feature weight (represented by the figure 1.), whereas in the actual analysis it got exceeded by three other types of introductors, is not significative, since the difference in the results seems mostly due to the relevance of the variables considered.

Grammatical subject	implicit	postposed	anteposed	total
Ortega y Gasset	836(30,0%)	572(20,6%)	1375(49,4%)	2783
Unamuno	5139(58,9%)	1730(19,9%)	1844(21,2%)	8713

Table 4. Classification of the corpora in sentences without overt subject, sentences with postposed subject and sentences with anteposed subject.

I would like to conclude with a global comparison of the prose and poetry corpora considered. Needless to apply a chi-square to the figures of table 4 in order to measure the significance of equivalences and discrepancies. On the one hand, both in prose and in poetry, almost one fifth of the sentences has an overt subject in postposition. On the other hand, another fifth of the sentences has the grammatical subject anteposed in poetry, whereas up to half of the sentences has it anteposed in prose. It might be the case that this significant frequency difference of subject anteposition is characteristic both for the literary genre and for the style of the author. More investigation is needed in order to check whether the difference is due to the genre rather than the author, or vice versa.

As for now, it can undoubtedly be stated that for both corpora, the one prosaic, the other poetic, the following features have a measurable effect upon the subject position : the sentence length, the type of verb and the opening of the sentence.

NOTES AND REFERENCES

1 ORTEGA Y GASSET, José (1927) : "Espíritu de la Letra", in : Obras Completas de José Ortega y Gasset (1966) Tomo III. Madrid : Revista de Occidente. UNAMUNO, Miguel de, 206 newspaper articles (1931-1936) published in GARCIA BLANCO (ed.) (1966) : Obras Completas : Miguel de Unamuno. Madrid : Escelicer. This corpus has been computerized by Prof. J. De Kock (Louvain).

2 CONTRERAS, Heles (1976) : A Theory of Word Order with Special Reference to Spanish. Amsterdam : North-Holland.

3 UNAMUNO, Miguel de (1928-1936) : "Cancionero", in : GARCIA BLANCO (ed.) (1964) : Obras Completas Tomo VI. Madrid : Escelicer.

4 DELBECQUE, Nicole (1974) : La colocación del regente con los verbos pronominales en español, and VAN DEN EYNDE, Els (1974) : La posposición del regente en español. Memoires, University of Louvain.

L'INDIGENO E L'ALLOGLOTTO NELL'ORDINE DELLE PAROLE DI UN IDIOMA DI FRONTIERA: L'ISTROROMANZO

Pavao Tekavčić

1. L'ordine delle parole (OP) è un dominio sintattico in cui, oltre ai fattori più specificamente linguistici, agiscono anche quelli stilistici: affettività, espressività ecc. D'altra parte, l'OP è anche strettamente collegato con la linguistica del testo (e del contesto), indirizzo conosciuto soprattutto sotto il nome tedesco di Textlinguistik. Mentre questi due campi di ricerche riuniscono quelli che potrebbero essere definiti i lati intra-linguistici del problema dell'OP, nel caso concreto del quale vogliamo dare qui un breve panorama vi si aggiungono anche i fattori extra-linguistici: i possibili influssi tra due idiomi vicini ma appartenenti a famiglie linguistiche diverse. L'idioma a cui è dedicata la presente comunicazione, l'istroromanzo, è infatti situato sulla cerniera slavo-romanza, nell'Istria sud-occidentale. Per il momento non disponiamo di studi sistematici sull'OP nei diversi dialetti veneti dell'Istria, ma dato lo stretto parallelismo e l'affinità tra l'istroromanzo e il veneto, crediamo di poter affermare che anche nel veneto istriano ci dovrebbero essere i medesimi influssi. Le ricerche future, dedicate specialmente a quest'ordine di problemi, potranno certamente portare alla luce diversi nuovi casi interessanti.

2. Siccome i dialetti istroromanzi sono idiomi di ceppo neolatino, vi si possono prevedere le caratteristiche generali della collocazione delle parole romanza, sulla quale perciò possiamo essere brevi.

2.1. In seguito alla perdita della declinazione sintetica si è avuta la funzionalizzazione sintattica della posizione reciproca del SN_1 (soggetto) e SN_2 (oggetto), e precisamente non soltanto nei casi in cui i tratti semantici (ad esempio

[+umano] per SN_1 e SN_2) permettono un'eventuale confusione (1), ma anche là dove essa è logicamente esclusa (2):[1]

(1) Lorenso vido Marowsa.	≠	Marowsa vido Lorenso.
(2) Al paron al liga al samer.	≠	Al samer al liga al paron.

L'ultima proposizione è grammaticale, ma semanticamente inaccettabile.

2.2. Se bisogna spostare il SN_2, l'OP si serve della trasformazione enfatica mediante gli elementi <u>a zi</u> - <u>ke</u>:

(3) A zi kwisti i zogatoli ke ti ge dagi a to feyo ?

La trasformazione enfatica non è tuttavia generale, perché in alcuni casi (in cui concorrono l'accento e l'intonazione) si ha la sola prolessi del SN_2:

(4) La kukera i vulí kavá ?
(5) Kwista karno i voy, meyo.

È da rilevare tuttavia che in questi casi il predicato è alla persona che esclude ogni possibile confusione.

2.3. La prolessi del SN_2 accompagnata dalla ripresa con un sostituente atono non serve a topicalizzare il SN_2 ma il V:

(6) Marowsa, Lorenso la vido.

È così anche nel caso della posposizione del SN_2 o del SP alla fine della proposizione:

(7) Al ge vuriva ben a sta mureda.

2.4. Gli elementi funzionali (preposizioni, congiunzioni) precedono sempre gli elementi retti da essi, come in genere nella sintassi romanza. Sarebbe banale citare degli esempi.

2.5. Per quanto riguarda la collocazione reciproca delle parti di un periodo, non ci sono differenze di rilievo tra l'OP istroromanzo e quello di altri linguaggi neolatini:

a) La proposizione temporale di solito precede la principale, ma può anche seguirla:

(8) Kwando k'al yo kapéy, al ge deys.

(9) Al zi livá kwando k'a yera ankura skowro.

La differenza nell'ordine reciproco delle due parti dipende dalle esigenze della topicalizzazione.

b) Anche la proposizione causale può precedere o seguire la principale:

(10) Sikume ka la kaza la ga baliva, al nu s'wo fidá da inpisá la lowme a pitrwolĝo... (esempio rovignese),

(11) ... a beña masáli de noto, parkí i maña de noto (sc. (gli insetti nella vigna).

c) Nelle stesse due posizioni può trovarsi anche una proposizione finale:

(12) Par disfantá el temporal i nostri veči i bruzava un po de yerba benedita...

(13) ... kwalkedown i li metiva fora anke per fáseli vidi, sti koverturi.

d) Nel periodo ipotetico ritroviamo la stessa duplicità, e valga come unico esempio il seguente, che riunisce entrambe le posizioni:

(14) Maestá, se vuy me dí la parola da no fa ñente s'i ve fagi vidi owna roba, i ve mostraré sta roba.

	Protasi:	Apodosi:
I	se vuy me dí la parola	i ve mostraré sta roba.
	Apodosi:	Protasi:
II	da no fa ñente	s'i ve fagi vidi owna roba.

3. Più interessanti saranno senza dubbio i casi che maggiormente si scostano dalla sintassi romanza "normale" e che sono dovuti a determinate evoluzioni morfosintattiche, peculiari dei dialetti altoitaliani (nel nostro caso più precisamente veneti) e dell'istroromanzo. Pensiamo anzitutto alla nota diminuzione dell'autonomia delle parole, risultato di più processi (perdita della declinazione, riduzione generale delle parti finali delle parole variabili, tendenza all'analiticità, cioè prevalenza dei mezzi sintagmatici su quelli paradigmati-

ci ecc.). Una conseguenza di quest'evoluzione sono le forme atone del sostituente personale che accompagnano obbligatoriamente il verbo; un'altra conseguenza, su cui ci soffermeremo un po' di più, è la particella a, che appare in due funzioni strettamente imparentate: essa funziona da SN_1 nei verbi che non ammettono un agente ("impersonali" o "unipersonali") (a) e anticipa il SN_1 (con altri verbi) se esso è invertito. Esempi:

a) a zi viro, a pyow, a me par impuseybile, a nu zi yowsto ecc.,

b) A zi sta own dey k'a se yo katá kwatro zermani, A se meto a maňá anke Bara Búrtolo (e centinaia di altri esempi).

Del tutto analoghi sono gli esempi rovignesi: a zi nwoto, skowro da lowna, a ğira frido, a ğira lá own vyečo ka paskiva su li grwote, ecc.[2]

La particella a non va confusa con il sostituente personale al, che spesso accompagna il verbo anche se il SN_1 precede immediatamente e che nell'eloquio rapido può essere realizzato come /a/; ad esempio:

(15) Al muré al se viseyna un po...
(nel parlare rapido: /almuréaseviséynaumpó/.

La differenza risalta beninteso nettamente nel femminile o nel plurale:

(16) La mureda la ven dowta varguňuza...
(17) Kwando ke i kwatro ameygi i zi zeydi vi...

La vocale del relativo ke e dell'omofona congiunzione si elide davanti alla particella a sicché ne nasce la sequenza /ka/, in cui la particella è tuttavia presente, come rivelato da semplici procedimenti trasformazionali:

(18) I te deygi k'a nu zi viro. = I te deygi: a nu zi viro.
(19) Kwil k'a yera lá, al ge dey. = A yera lá owno, e al ge dey.

Visto che l'apertura della /e/ protonica in /a/ è sporadicamente presente in tutti i dialetti istroromanzi, ed è massiccia appunto nel rovignese, l'etimologia della particella <u>a</u> non presenta problemi: essa è geneticamente identica all'italiano antico <u>egli</u>, <u>ei</u>, <u>e'</u> e al francese <u>il</u> (cf. ad esempio <u>a zi viro</u> - <u>egli è vero</u> - <u>il est vrai</u>) e si riconnette al sostituente tardolatino ILLU.[3]

4. Dalle ricerche da noi condotte in precedenza e pubblicate in altra sede[4] risulta assai bene rappresentato e tipico per l'istroromanzo il non-accordo del participio (parte del SV) con il SN posposto, e precisamente non soltanto SN_2 - come è comune nelle lingue letterarie della Romània centrale - ma anche SN_1. L'esempio-guida per quest'ultimo caso può essere questo:

(20) A zi viñów i preti. È/Sono venuto i preti.

Va precisato che il non-accordo concerne soltanto le persone dei non-interlocutori (la 3ª e la 6ª) e, all'interno di esse, interessa solo il participio, non l'ausiliare, giacché nessun verbo in nessun paradigma di qualsiasi dialetto istroromanzo distingue formalmente la 3ª dalla 6ª persona. La forma <u>zi</u> vale dunque tanto 'è' quanto 'sono' (essi).

Non sarà privo di interesse confrontare l'accordo istroromanzo con quello dell'italiano letterario:

	Istroromanzo:	
SN_1 + SV:	<u>I preti zi viñówdi</u>.	Accordo: +
SV + SN_1:	<u>A zi viñów i preti</u>.	Accordo: -
SN_2 + SV:	<u>La kaza i (la) ye vindowda</u>.	Accordo: +
SV + SN_2:	<u>I ye vindów la kaza</u>.	Accordo: -

Nell'italiano letterario, al contrario, si ha l'accordo in tutti e tre i primi casi, mentre il non-accordo è limitato normalmente al quarto caso (SV + SN_2).[5]

I dati statistici per i due maggiori dialetti istroromanzi fanno vedere una netta prevalenza del non-accordo sull'accordo: nei materiali dignanesi da noi esaminati si hanno 25 casi di non-accordo contro 6 casi di accordo (approssimativamente

4 : 1), mentre nei materiali rovignesi la proporzione è di 53 casi di non-accordo contro 19 casi di accordo (su per giù 3 : 1). Dato lo strettissimo parallelismo fra le due varietà dell'istroromanzo, la differenza tra le proporzioni non può essere dovuta a cause di ordine linguistico, ma sarà con ogni probabilità in relazione al genere del materiale di cui ci siamo serviti. Mentre per il dignanese - essendo stati pubblicati meno testi in questa varietà - abbiamo lavorato principalmente su testi di carattere spontaneo, da noi raccolti in loco, per il rovignese, il quale vanta molti più testi pubblicati, abbiamo usato precipuamente un corpus costituito da testi scritti.[6]

Il non-accordo citato non è ignoto nel dominio veneto, mentre è pressoché assente dai testi friulani: infatti, nel materiale friulano da noi esaminato la proporzione tra l'accordo e il non-accordo va da 100 : 0 (nei testi pubblicati nelle annate spogliate della rivista Sot la nape) fino a 78 : 22 (nelle Reliquie ladine pubblicate da J. Cavalli). Riunendo tutti i risultati, la proporzione viene ad essere all'incirca questa: 90 % di casi di accordo contro 10 % di casi di non-accordo.[7] Il friulano è dunque praticamente escluso come fonte primaria o secondaria del non-accordo istroromanzo. Anzi, i pochi casi di non-accordo nei testi friulani saranno dovuti ad influssi veneti, forse anche istroromanzi: probabilmente non è casuale che il maggiore numero dei casi di non-accordo si abbia nel tergestino e nel muggisano, cioè nelle due varietà friulane più lontane dal nucleo friulano e più "immerse" nell'ambiente veneto.

Dagli spogli risulta una chiara tendenza a far dipendere l'accordo del participio dall'OP, più precisamente dalla collocazione reciproca del SN e del SV, qualunque sia la funzione del SN (soggetto-oggetto) e qualunque dei due ausiliari venga adoperato: l'accordo tende ad essere riservato alla sequenza SN - SV ed eliminato dalla sequenza SV - SN tanto con 'essere' quanto con 'avere'. Il participio in tal caso assume la forma non marcata delle categorie del genere e del numero (maschile singolare), che è come provvisoria, mentre il SN_1 posposto aggiunge la necessaria precisazione. Senza poter per adesso constatare in che modo, siamo del parere che l'altissima percentuale dell'accordo e la distinzione formale tra la 3ª e la 6ª persona nel friulano da un lato, la prevalenza del non-accordo e l'identità totale della 3ª e della 6ª persona nell'istroromanzo (e in misura un po' minore anche nel veneto) dall'altro, siano reciprocamente collegate.

5. Mentre i fenomeni finora esaminati possono essere di origine puramente neolatina, dunque interna, la caratteristica di cui intendiamo parlare in questo paragrafo non si spiega nell'àmbito della sintassi degli idiomi romanzi, almeno per quanto concerne il nostro caso. Si tratta dell'inserzione di un avverbio (o una parola funzionalmente affine) tra il sostituente personale atono e il verbo, collocazione normalmente impossibile nelle lingue romanze. Ecco alcuni esempi:[8]

(21) Mey i me magari vergóñi un po.
(22) La lo kwazi impinéyso.
(23) ...la se un po vergóña...
(24) La bala sanpátika la va aturno, ke la nu se ñanka vido.[9]
(25) ... i nu ma ñanka rakurdívo. (esempio rovignese)
(26) A nu sa ñanka mana stasira? (" ")[10]

Come rilevato nei nostri lavori citati, tali inserzioni sono rarissime nell'italiano antico (e limitate soltanto a <u>pure</u>) e non sono frequenti nemmenọ altrove (alcuni casi ci sono nelle lingue iberoromanze). È ovvio dunque che questi idiomi non possono essere la fonte dell'inserzione nell'istroromanzo. Inoltre, va sottolineato che gli esempi istroromanzi si trovano nei due maggiori dialetti (e con somma probabilità anche negli altri), nel materiale antico e contemporaneo, nei dialoghi del tutto spontanei e nei testi redatti e stampati. Questi fatti provano che l'inserzione non è un fenomeno marginale ma costante ed integrato nel sistema. L'unica spiegazione possibile del fenomeno istroromanzo è l'influsso dell'adstrato slavo (croato, sloveno), nel quale l'inserzione analoga è non soltanto possibile ma addirittura normale e regolare. Basta confrontare gli esempi istroromanzi con le relative traduzioni in serbocroato:

(21a) Ja se možda malo stidim.
(22a) Ona ga gotovo napuni.
(23a) ...ona se malo stidi...
(24a) Kuglica se vrti uokolo (tako), da se niti ne vidi.
(25a) ...nisam se ni sjetio.
(26a) Zar se uopće ne jede večeras?

6. Siccome gli idiomi slavi dell'Istria conservano la declinazione sintetica e, grazie a questo, anche una notevole autonomia delle parole e libertà dell'OP, crediamo che un influsso slavo non sia del tutto estraneo neanche in certe altre particolarità della collocazione delle parole, come ad esempio la prolessi di un avverbio (o parola funzionalmente equivalente) della quale ci sono alcuni esempi nei materiali sia antichi che contemporanei. Valgano i due esempi seguenti: il primo desunto dalla versione rovignese della Parabola del Figliol Prodigo, risalente al 1835 e fornita forse da P. Stancovich,[11] il secondo tratto da un aneddoto in dignanese, da noi registrato a Dignano nel 1963. Va rilevato che nessuno dei due testi è peraltro caratterizzato da una speciale affettività o spontaneità.

(27) E daspoy no poki di, dowto ingrumá el feyo pyun zúveno via dal país al ze zi a largo purasé.[12]
(28) ...e 'l barba ge da pyown soldi de l'altra volta.... e low, alura,... pyown dey fora al zi sta.

Nel primo esempio troviamo due inversioni: le parole <u>via dal país</u> sono il determinante (spaziale) del verbo <u>al ze zi</u>, e la parola <u>purasé</u> è il determinante (intensificante) di <u>a largo</u>. L'OP "normale" sarebbe invertito in ambedue i casi: <u>al ze zi via dal país purasé a largo</u>.

Il secondo esempio (28), nella sua forma piena, rientra anche nel fenomeno di cui ci occupiamo nel paragrafo seguente.

7. È noto e diffuso il costrutto che assume la forma di un

chiasmo, risultato della sovrapposizione di due ordini contrari: a - b e b - a dànno a - b - a. Esempio tipico: ho bevuto un litro ho bevuto, risultato della sovrapposizione di ho bevuto un litro e un litro ho bevuto. Il primo ordine è affettivamente neutro, il secondo è marcato perché presenta la prolessi del SN_2. Ecco alcuni esempi raggranellati nei nostri testi:

(29) Despoy...i yeri spuzá preyma da zey in Amerika i yeri spuzá.

(30) ... al čo i kwatro soldi e li meto own zura l'altro li meto.

In qualche esempio al posto del verbo da ripetersi appare un sinonimo:

(31) ... e a pol ési anka 'l kuntraryo a pol viñéy... (si tratta di risultati di un gioco),

(32) La kambowza zi el depozito de víveri se čama kambowza.

Più numerosi sono pur sempre i casi in cui le due proposizioni rimangono complete (a - b - b - a):

(33) I ye navigá ankúra, ankúra i ye navigá.

(34) I veyn fato veyntiseynkwe dey de mankansa, veyntiseynkwe dey de mankansa i veyn fato.

(35) I vivi duy lavandere par lavá, par lavá...i vivi duy lavandere.

(36) ...aturno al fugulér a yera pyen de zento, pyen de zento a yera.

(37) ...al zi sta pyown dey fora, pyown dey fora al zi sta. (cf. sopra 28).

(38) ...la yera povera, me muyír, ah, povera la yera.

Gli esempi 33-37 provengono da un informatore che ci ha fornito più testi di tutti gli altri e che conosce e capisce la problematica degli studi istroromanzi meglio degli altri; ma proprio quest' informatore ha anche più cultura degli altri e più conoscenze alloglotte, e ha pure viaggiato molto durante tutta la sua vita. I testi avuti da lui non sono mai registrazioni di discorsi e dialoghi spontanei, ma brani raccontati ad un ritmo piuttosto lento, che consentiva riflessioni e scelte di termini e costrutti. L'esempio 38 è dovuto invece ad un altro informatore, che parlava assai presto e in modo molto concitato, dunque del tutto spontaneo.

Di conseguenza, anche il chiasmo citato si trova in materiali di diverso genere.

8. Ecco infine due casi di inversione delle parti del periodo temporale tipo sono due giorni che... In tali casi la proposizione formalmente principale è in verità funzionalmente dipendente, perché determina l'altra la quale è formalmente dipendente (sono due giorni che non lo vedo = non lo vedo da due giorni). I materiali istroromanzi offrono alcuni esempi del costrutto contrario, nel quale la proposizione determinante è anche formalmente dipendente ed è introdotta da /ke/. Ecco due esempi:

(39) Mey ye le maróyde k'a zi ğize ani.
(40) Adeso i soyn in pensyón k'a zi sete ani.

9. La rassegna precedente è soltanto un breve panorama delle caratteristiche dell'OP istroromanzo, che potrà essere ampliato e completato da ricerche più approfondite. Gli esempi, tolti dai due principali dialetti, da testi di genere e di periodi diversi e ottenuti anche da informatori assai diversi, conferiscono ai dati un valore rappresentativo dell'intero dominio. I tre ordini di fatti riscontrati (tratti romanzi generali, tratti altoitaliani e veneto-istroromanzi, infine tratti alloglotti dovuti all'influsso slavo) ci hanno permesso di individuare ciò che dice il titolo: l'elemento indigeno e l'elemento alloglotto nell'ordine delle parole dell'istroromanzo.

NOTE

1 Salvo indicazioni contrarie, vengono dati esempi dignanesi. Tutti gli esempi sono stati conformati al sistema grafico da noi usato in altri lavori. Gli accenti vengono segnalati soltanto nei casi in cui la loro posizione potrebbe dare luogo a dubbi.

2 Cf. DEANOVIĆ 1954, pp. risp. 72,71,92.

3 Cf. TEKAVČIĆ 1967a, pp.279-280; 1976.

4 Cf. TEKAVČIĆ 1967b.

5 Cf. TEKAVČIĆ 1972, paragrafi 1175-1177.

6 Per gli altri dialetti istroromanzi trattati nel nostro studio (TEKAVČIĆ 1967b) ci siamo potuti valere unicamente dei testi pubblicati da A.Ive (IVE 1900).

7 Cf. TEKAVČIĆ 1967b, pp. 177-178.

8 Cf. TEKAVČIĆ 1969, pp. 301-303.

9 Cf. IVE 1900, p.194.

10 Cf. DEVESCOVI 1894, pp. 49 e 53.

11 Cf. DEANOVIĆ 1954, p.55.

12 Il senso e il parallelismo con le altre versioni permettono di correggere daspoy no poki di sia in daspoy no molti di che in daspoy poki di.

BIBLIOGRAFIA

DEANOVIĆ, Mirko (1954): Avviamento allo studio del dialetto di Rovigno d'Istria. Zagreb: Sveučilište u Zagrebu.

DEVESCOVI, Raimondo (1894): Vita rovignese. Bozzetti in vernacolo. Rovigno.

IVE, Antonio (1900): I dialetti ladino-veneti dell'Istria. Strasbourg.

TEKAVČIĆ, Pavao (1967a): "Današnji istroromanski dijalekt Vodnjana". Rad Jugoslavenske akademije znanosti i umjetnosti 348: 141-288.

--- (1967b): "Il non-accordo tra il predicato composto con 'essere' ed il soggetto invertito nei dialetti istroromanzi". Studia Romanica et Anglica Zagrabiensia 23: 161 - 194.

--- (1969): "Testi istroromanzi dignanesi". Travaux de Linguistique et de Littérature publiés par le Centre de Philologie et de Littératures Romanes de l'Université de Strasbourg VII: 275-303.

--- (1972): Grammatica storica dell'italiano.Bologna:il Mulino.

--- (1976): "Historijske komponente istroromanske sintakse". Comunicazione alla Conferenza internazionale sulla sintassi storica, Kraków 22-23 marzo 1976 (in corso di stampa negli Atti).

THE STRUCTURE OF MORPHOLOGICAL FAMILIES

H. Joachim Neuhaus

To Marvin Spevack, on his 50th birthday

1. Language typology has been morphological typology until quite recently, when syntactic questions of word order and word order change became more prominent in typological studies. But whereas any natural language has a syntactic structure, not every language has a developed morphology as a separate word-formation system. This is why morphological properties are per se language specific and not universal. As the longwinded 19th century debate about morphological types has shown, nearly all languages - if they use word-formation systems at all - employ various methods or structural types simultaneously. This makes for a greater variety and complexity of structural relations, and lends a certain plausibility to the well-known dictum of BLOOMFIELD (1933: 207):

> .., languages differ more in morphology than in syntax. The variety is so great that no simple scheme will classify languages as to their morphology.

1.1. Some typologists have tackled this profuse situation statistically. There have been several suggestions for statistical indices (e.g. GREENBERG 1954a, KRUPA 1965). But for the purpose of typological differentiation these indices have turned out to be of rather limited value. Calculations from running texts show a relatively homogeneous behaviour for different languages.

To give one example: a 2.2 ratio of morphemes to words in text material ('index of analysis') from Yokuts, a Californian Indian language, is according to a remark by GREENBERG (1954b: 200):

> ... just about the same as for all other languages (to be sure, a relatively small group) that I have examined. In this respect, then, Yokuts is normal.

The impression of normality with statistical indices is, of course, due to the sampling method. Quite different from syntactic constructions, the potential complexity of morphological structures is very much reduced in actual language use.

Working with Yokuts informants, NEWMAN (1954: 85) reported this phenomenon, and similar experiences are obvious with other languages as well:

> As part of my field procedure I invented many such complex words, which my informant dutifully translated, though with obvious amusement. These heavily suffixed words were grammatically unobjectionable. But they were creations of a structurally oriented linguist, not those of native speakers who collectively impose selective limits upon the structural potentialities of their language.

1.2. Even less satisfying results are obtained if informants are encouraged to venture intuitive opinions on morphological matters. Again there has been a pertinent observation from BLOOMFIELD (1933: 208):

> One soon learns that one cannot look to the speaker for an answer, since they do not practice morphologic analysis; if one bothers them with such questions, they give inconsistent or silly answers.

1.3. What makes the situation even more complicated is the familiar phenomenon of a language-dependant analytic capability. Analytic attitudes towards word-formations vary considerably even between closely related languages such as English and German. SAPIR (1921: 195) made some interesting comments on these differences:

> Such words as _credible_, _certitude_, _intangible_ are entirely welcome in English because each represents a unitary, well-nuanced idea and because their formal analysis (_cred-ible_, _cert-itude_, _in-tang-ible_) is not a necessary act of the unconscious mind (_cred-_, _cert-_, and _tang-_ have no real existence in English comparable to that of _good-_ in _goodness_). ... In German, however, polysyllabic words strive to analyse themselves into significant elements. ... Latin-German words like _kredibel_ "credible" and French-German words like _reüssieren_ "to succeed" offer nothing that the unconscious mind could assimilate to its customary method of feeling and handling words. It is as though this unconscious mind said:"I am perfectly willing to accept _kredibel_ if you will just tell me what you mean by _kred-_."

All these empirical data are still poorly understood from a psycholinguistic point of view, and very few systematic experimental investigations have dealt with this problem.

1.4. Given these general observations one has to expect that the study of morphological typology should turn out to be rather abstract and inaccessible. Abstract, because underlying systems have to be reconstructed as highly complex structural resources with only relatively slender use of these resources as empirical evidence. This is why prima facie evidence has quite often led to oversimplifications in morphological descriptions. Inaccessible, because crucial empirical evidence is hard to get and successful data preparation involves huge quantities of morphological constructions. Computer files with more than 200,000 word entries had to be searched and analysed in order to be able to demonstrate the non-planarity of morphotactic cycles in English suffixation (cf. NEUHAUS 1975).

2. For a long time the standard pattern for the structural description of word-formation systems has been a linear ordering theory of morpheme classes. This linear ordering is implicit in HYMES' 'positional analysis' (1955), or in STOCKWELL's terminological suggestion of 'morphotactics' (cf. HILL 1958: 68). Nearly all work with so-called polysynthetic languages still uses this frame, and there is a tendency to think it to be adequate in general. It can also be found in more recent works on English and French that use 'primary and secondary' derivation theories (e.g. STEIN 1971). The linearity condition was strong enough to lead CHAPIN (1970) in his M.I.T. dissertation to postulate an epicyclic ordering.

2.1. In the following there will be arguments for the abandonment of linear morphotactics, which is an empirical matter, and arguments for the adoption of a graph-theoretic framework, which is a theoretical suggestion. Graph-theory lends itself to a rigorous definition of somewhat intuitive structural concepts such as 'morphological family', but more important, there are quite useful graph-theoretic properties that can enter into complexity measurements and typological subgroupings (cf. NEUHAUS, forthcoming a).

What it means to use a graph-theoretic framework can be explained quite easily, without too many technicalities. English prefix-formations, constructions such as undecomposed, irreplaceable, etc., can be used for a nonlinearity proof. Figure (1) gives an incidence matrix for some prefixes. This is meant as

><							
0•	un-						} strictly ordered
10	0•	in^2-					
11	10	1•	re-				} reduplicating
11	11	10	1•	sub-			
10	00	10	10	0•	en-		} terminal
10	00	10	00	00	0•	be-	
><	><	><	><	><	><		

Figure (1) Incidence matrix for six English prefixes

a minimal notation to show their respective morphotactic interrelationship. A corresponding graph - a good drawing of an essentially planar graph - is given as Figure (2). The prefixes un- and in^2- are shown to be strictly ordered. The '10' notation in the respective cell of the incidence matrix allows for formations such as: uninfallability ('1 ') but excludes a reverse pattern: *in-un- (' 0'). This is a purely morphotactic constraint There are no semantic arguments that could be used since both prefixes coincide semantically in their strong tendency to denote contrary opposition.[1] Both in^2- and un- do not reduplicate: '0•'. There is just one un-un- formation we know of and that is used in a textbook (HILL 1958: 137). It should have earned a star by now. Reduplication ('1•') is possible with re- and sub- as well as with a few other prefixes: sub-subdivision, sub-subcontract, etc.. Terminal prefixes such as en- and be- close possible prefix strings. They take the last position. Given the matrix notation only zero values (' 0') are admissible as second digits.

2.2. There are, of course, many other morphotactic properties that can be studied in detail once morphological data have been prepared in this graph-theoretic framework. A most interesting structural property of all English prefixes is their systematic interrelation.[2] Together they form a totally connected graph. Given a good drawing of the English prefix system there is thus no isolated prefix, but there are three structurally different

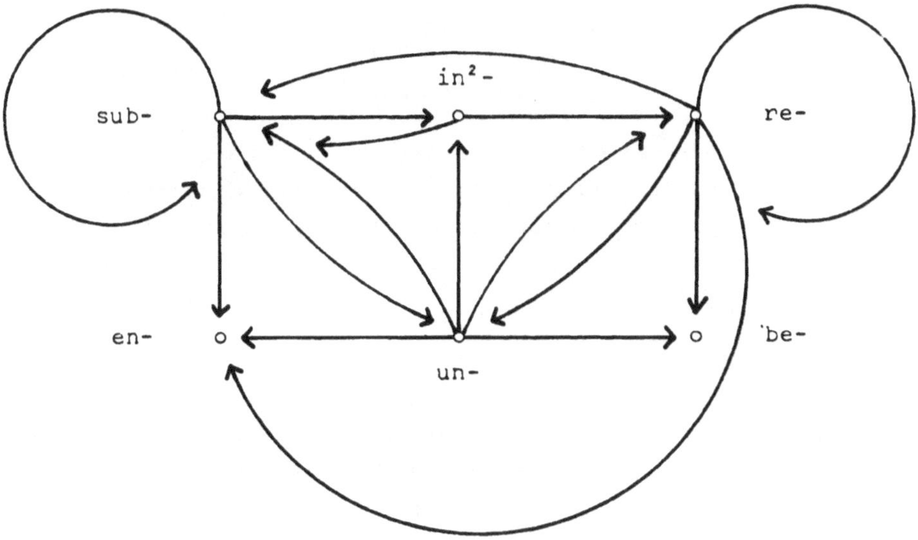

Figure (2) Graph for six English prefixes

classes of prefixes: cyclic prefixes, which have a set of admissible immediate predecessors as well as a set of admissible immediate successors, and initial and terminal prefixes, which have only one of these sets each:

Cyclic prefixes	Initial prefixes	Terminal prefixes
{ } Π { }	Π { }	{ } Π

2.3. The complexity of a system can either be studied from an overall perspective or from the point of view of one of its elements. The notion 'morphological family' belongs to the latter type. The immediate prefix neighbours of a prefix may be said to form its core family. If examples from the fragment of Figure (2) are used the following core family for <u>re</u>- emerges:

predecessors: u n - r e - movable
i r - r e - claimable

successors: r e - u n - fold
r e - s u b - merge
r e - b e - hold
r e - e n - list

Furthermore there are reduplicating cases as r e - r e - form, where a systematic phonological differentiation between a weak and a strong prefix realization occurs. Of course, that also has semantic implications, but it does not affect the underlying morphotactic regularities.

2.4. Prefixes may be ordered according to core family size. Differences in numerical complication can then be studied and indices for the degree of connectivity of an overall system can be calculated and compared. A rather straightforward presentation may use matrices such as Figure (3) for this purpose.

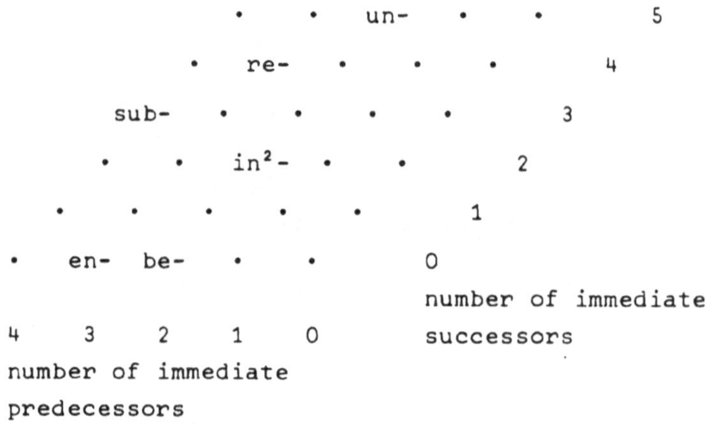

Figure (3) Connectivity profile for six English prefixes

2.5. In the larger framework of typological comparisons structural descriptions of this type will lead to insights into the complexity of morphological systems, and it should be more meaningful to have a typology of underlying systems than to have mere statistics on the average use of the resources available in texts.

NOTES

1 Here, as in the following only morphotactic well-formedness characteristics are dealt with. On the relation between morphotactic constraints and semantic and phonological constraints cf. NEUHAUS (1975) and (forthcoming b).
in^2- is the negating prefix (invariant, illegitimate, impolite, etc.).

2 A more comprehensive description of the evidence available for this structural discovery is in progress.

References

BLOOMFIELD, Leonard (1933): Language. Chicago.

CHAPIN, Paul (1970): "On affixation in English". BIERWISCH, Manfred/HEIDOLPH, Karl (eds.): Progress in linguistics. The Hague. 51-63.

GREENBERG, Joseph (1954a): "A quantitative approach to the morphological typology of language". SPENCER, R. (ed.): Method and perspective in anthropology. Papers in honor of Wilson de Wallis. Minneapolis. 192-220.

--- (1954b): HOIJER (ed.): 200.

HILL, Archibald (1958): Linguistic structures. From sound to sentence in English. New York.

HOIJER, H. (ed.) (1954): Language in culture. Conference on the interrelations of language and other aspects of culture. Chicago.

HYMES, D. H. (1955): "Positional analysis of categories. A frame for reconstruction". Word 11: 10-23.

KRUPA, Viktor (1965): "On quantification of typology". Linguistics 12: 31-36.

NEUHAUS, H. Joachim (1975): "Morphotaktische Zyklen". RIX, Helmut (ed.): Flexion und Wortbildung. Akten der V. Fachtagung der Indogermanischen Gesellschaft, Regensburg, 9.-14. September 1973. Wiesbaden. 220-231.

--- (forthcoming a): "Morphotactical complexity". CHATMAN, S./ ECO, U. (eds.): Collected papers of the first congress of the International Association of Semiotic Studies. The Hague

--- (forthcoming b): "Wortbildungssemantik". BREKLE, E./ KASTOVSKY, D. (eds.): Festschrift für Hans Marchand zum 70. Geburtstag. Tübingen.

NEWMAN, Stanley (1954): "Semantic problems in grammatical systems and lexemes: a search for method". HOIJER (ed.): 82-91.

SAPIR, Edward (1921): Language. An introduction to the study of speech. New York.

STEIN, Gabriele (1971): Primäre und sekundäre Adjektive im Französischen und Englischen. Tübingen.

2. SYNTAX UND GENERATIVE THEORIE

ON A NOT SO PERFECTLY ARGUED DESCRIPTION OF THE PRESENT AND THE PRESENT PERFECT IN ENGLISH

Heinz W. Viethen

In her dissertation and in a further contribution, Mary Jane GALLAGHER (1970a and 1970b) deals - among other things - with problems of tense in English and in particular, with characteristic features of the underlying structures of the Present and the Present Perfect. - In my paper, some aspects of her investigation are outlined and examined critically; finally, some suggestions for a somewhat different approach will be advanced.*

1. Within the framework of an analysis of _have_ and the Perfect in English, Mary GALLAGHER (1970b: 63) reaches the following conclusions (among others):

> First, _Tense_ need not be filled in the deep structure: some Verbs must be _tenseless_ [: statives and habituals are those Verbs that have no Tense in the deep structure (1970b: 63)]. Second, there is no need to generate both Adverbs of Time and Tense, since the latter is predictable from the former, but not conversely.

These two claims - that statives and habituals are tenseless and that tense is predictable from adverbs of time - are closely linked, and I would like to have a closer look at Gallagher's justification of them.

1.1. It was Chomsky who reported a suggestion by Barbara Hall [Partee] that

> each element of the Auxiliary has associated with it certain characteristic Adverbials that may (or, in the case of Present tense, must) co-occur with this Auxiliary element, ... (CHOMSKY 1965: 216, fn. 23).

Taking this suggestion to mean that the presence of time adverbs be dependent on _Tense_, it follows for Mary Gallagher that "_Tense_ [must] be obligatory wherever time adverbs are allowed" (1970a: 220). Assuming furthermore that _right now_ is the proper adverbial for the element _Present_, M. Gallagher concludes:

> If _Present_ chose _right now_ and only _right now_ or its synonyms, we should expect 1) and 2) to be synonymous on a reading, which they are.
> (1) John is signing the declaration of Independence.
> (2) John is signing the declaration of Independence right now.
> (GALLAGHER 1970a: 220f.)

Not synonymous, however, are sentences like the statives mentioned under (3), or like the habituals under (4) (ex. taken from GALLAGHER 1970a: 221):

(3a) John is a man.

(3b) John is a man right now.

(4a) John works hard.

(4b) John works hard right now.

The difference between the first and the second member of each pair is that between reference to an unchanging and to a temporary situation, respectively, with right now signalling "impending, or at least possible, change" (1970a: 221). - To cut a long story short, Mary Gallagher decides that it is generally preferable to base the choice of a particular tense on the occurrence of the appropriate adverb(s). Accordingly, sentences with what Mary Gallagher calls an "original" present tense, i.e. sentences with the deep structure element Present, must permit the presence of right now in surface structure. However, if the introduction of right now results in a change of meaning (as in (4b)) or in a fairly awkward sentence (as in (3b)), the respective verbs are assumed to be "without tense in underlying structure" (1970a: 222). The alternative approach - to stick to the procedure suggested in CHOMSKY 1965 and to delete right now, when it occurs together with statives or habituals - is not taken, as such a meaning-changing transformation is not considered to be desirable.

1.2. Before we mention other data which might support the choice of Gallagher's alternative, we would like to comment further on the choice of right now as an adverb of time to be associated with the element Present in deep structure. Clearly, Mary Gallagher did not read carefully that passage from CHOMSKY 1965, which provided her with the "Tense chooses Adverb"-approach and which we quoted above. Where our quotation stops, Chomsky goes on:

> and the characteristic Adverbials of P r o g r e s s i v e [my emphasis] do occur with the verbs own, understand, know, etc. (cf. "I know the answer r i g h t n o w [my emphasis]," alongside of "I know the answer"), although ... (CHOMSKY 1965: 216, fn. 23).

If Chomsky's classification of right now as "adverbial of progressive" is correct, it is small wonder that sentences like (3a) and (3b) or (4a) and (4b) cannot be synonymous, as statives and habituals can only take the Progressive under special and very restricted circumstances, as can easily be seen from Schopf's investigations, for example (cf. SCHOPF 1963 and 1969). It should be obvious that - taking right now - Gallagher chose the wrong adverb for proving her point. Of course, this conclusion is not sufficient to refute her basic argument, namely that statives and habituals have no tense in deep structure, as even the introduction of adverbs like at present or currently into sentences like (3a) or (4a) does not result in synonymy with these:[1]

(3c) *?John is a man at present.

(3d) *?John is a man currently.

(4c) *?John works hard at present.

(4d) *?John works hard currently.

1.3. As a further justification for the claim that there are sentences without tense-specification in deep structure, Mary Gallagher adduces sentences where the main verb cannot appear with a past time adverb, although the main clause as well as an embedded relative clause show the morphological indications of the past tense. One of these examples is (5):

(5) The man who answered the telephone was Henry's brother (*last week).

Due to the unacceptability of the adverbial last week, (5) cannot contain an "original" past. It also cannot be said to contain an original present (as defined by Gallagher), since a present time adverbial is not acceptable either, as can be seen from sentence (6):

(6) *The man who answered the telephone was Henry's brother at present.

The most convenient explanation is again to assume that the main clause, which is stative, is tenseless in deep structure and that the surface structure past ending on the main verb is due to the (deleted) past time adverb of the embedded relative clause (cf. (7a)):

(7a) The man who answered the telephone yesterday was Henry's brother.[2]

We might, therefore, be inclined to accept Gallagher's tenseless representation of statives in deep structure, if there were no sentences like (8a):

(8a) The woman who answered the phone was Henry's girlfriend last week.

It should become obvious from sentences like (8a) that it is not sufficient to postulate tenseless, and that means adverbless, deep-structure representations for statives, without indicating any further what kind of state is involved: whether it lasts for "(nearly) all time" as in generic sentences, cf.

(9) A horse is a mammal.

or whether the state is fairly permanent, at least after it has been achieved, as for example in (10a),

(10a) John's father is tall.

or whether it is not so permanent, as in

(11) Paul is angry.

One may of course wish to claim that such distinctions as might be involved in the states mentioned in (10a) and (11) are only based on our knowledge of the world: we just know that anger does not last very long, while tall people will never become smaller.[3] However, it is not only the case that (11) permits the addition of at present, while such an adverb would at least be felt to be very curious in the case of (10a); we must also point out that such differences which obtain with respect to the durability of states are relevant for the use and interpretation of the Progressive Form as well (cf. e.g. SCHOPF 1963).

1.4. That it is not completely sufficient to assume the existence of tenseless statives also becomes obvious from another example adduced by Mary Gallagher in order to prove her point, namely sentence (12):

(12) The foul-tempered dog hid under the table.

She does not wish to derive this sentence "either from structures with a verb in the relative clause that is an original present [as in (13a)] or from one that is an original past [as in (13b)]" (GALLAGHER 1970a: 223):

(13a) The dog that is foul-tempered hid under the table yesterday (when he was sweet-tempered and even timorous).

(13b) The dog that was foul-tempered hid under the table yesterday (and emerged, after some soul-searching, as the mild creature now before you).

Deriving (12) from either (13a) or (13b) does of course involve the theoretical difficulty that a meaning-bearing element is erased in Relative Clause Reduction (cf. GALLAGHER 1970a: 223), but on the other hand, assuming the (stative) relative clause to be without tense does not avoid that problem by any means! It only obscures the problem which consists in the fact that statives do not form a uniform class at all, as should become obvious from sentence (14), which is - superficially - entirely parallel to (12):

(14) The short-tailed dog hid under the table.

It is "knowledge of the world" again that dogs can be foul-tempered today and good-natured tomorrow, while short-tailed dogs remain short-tailed as long as they live. Nevertheless I should claim that only in the case of (14) can the embedded clause which undergoes Relative Clause Reduction justifiably be assumed to be tenseless. As opposed to (14), (12) must be derived from something like (13a) or (13b), in order to explain the ambiguity of the adjective in (12) which otherwise remains unrepresented.

1.5. So far, I have not said much about a justification of Mary Gallagher's assumption that temporal adverbs determine tense and that it cannot be done the other way round. That is due to the fact that I am not at all certain whether this alternative is really important. What is important, however, is her claim that certain sentences do not have tense in their deep-structure representation. Basically, that claim is justified, although - as indicated above - it is not sufficient to attribute tenselessness to all statives and habituals without any further qualifications.

Furthermore, we should say that it is impossible to avoid consideration of sentences like (15) and (16):

(15) The Incas were a highly civilized people. (BALD 1972: 62)

(16) Horses have always been mammals.

Clearly, these cannot be anything but tenseless in their deep structure, as they are not only stative, but also valid for "all time". On the other hand, they do not easily conform to Gallagher's rule of Present Insertion, which is supposed to account for the appearance of the morphological Present in all those tenseless and habitual sentences, where (surface) tense is not determined by the tense of a preceding clause. While the past tense in (15) might be derived from a deleted relative clause like <u>who</u> <u>existed</u> <u>from</u> ... <u>to</u> ... , such a solution is not open to sentences like (16). There, tense could only be derived from a matrix sentence like

(17) It Tense/Aspect (+Adverbial) be the case

with a t e n s e l e s s embedding of the sort <u>horses</u> <u>be</u> <u>mammals</u>. Whether such a tense-carrying deep-structure matrix could not also be employed for all statives and habituals which do not have tense in their own deep structure clause, might be a point for further investigation.

2. Statives do not easily take temporal adverbs, but they are not unique in this. The same applies for example to the English Present Perfect, which is also considered by Gallagher, and quite extensively, too. One of her basic claims is that the structures of possessive and present perfect sentences are not only connected through a continuous historical development, which is confirmed by English historical grammars, but that such a connection can also be established for present-day English. Her aim is "to analyse one type of periphrastic perfect" (GALLAGHER 1970b: 52), namely what she calls the "perfective" type and what others have referred to as Resultative Perfect:

> The procedure [is] to propose a deep structure in which there is a relationship between the periphrastic perfect and the possessive use of <u>have</u> and, in justifying that proposal, to indicate that the historian is incorrect [in his assumption that there is no continuing relationship between the possessive and the perfective <u>have</u>] (GALLAGHER 1970b: 52).

2.1. First of all we would like to reproduce the deep-structure tree (18b, cf. next page) provided for sentences like (18a),

(18a) John has finished that job.

although we will discuss only some of the characteristics of that tree and their justification.

(18b)

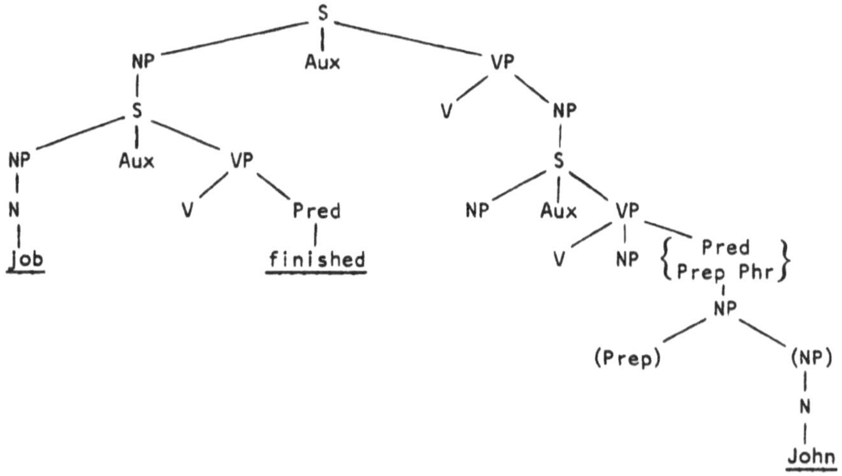

(GALLAGHER 1970b: 53)

It should become quite clear from the diagram that Mary Gallagher represents the "compound" nature of the Present Perfect by assuming a deep structure which contains two embedded sentences: one of them contains the "agent", while the other refers to the "result" of the action, i.e. to the fact that the job is finished.

2.1. Dealing with the S embedded on the left, her first claim is now

> that the embedding is a static passive. This is in accordance with the surface structure of the older form, in which the complement has the word order of the passive:
> [19] John has that job finished.
> [20] That job is finished. (GALLAGHER 1970b: 53)

She strongly emphasizes that a static and not a kinetic passive is involved. S t a t i c passives are incompatible with by-phrases indicating an agent and also with a "definite Time Adverb in the same clause". As an example we may refer to Gallagher's sentence (21),

(21) That toast is burnt (*by John) { (*at noon) / (*at present) }

which - in the "static" interpretation - neither admits an agent nor a temporal adverb, as long as we stick to the present tense. As it might be difficult to see the difference between what Gallagher calls a static and a kinetic passive, we might refer to German, where we find a similar differentiation in the sein- and the werden-passive: the werden-passive as in (22a)

(22a) Das Holz wird (von Christoph) verbrannt.

is "kinetic" and compatible with an agent, whereas the sein-passive as in sentence (22b) is incompatible with an agent:

(22b) Das Holz ist (*von Thomas) verbrannt.

Keeping these German examples in mind, the incompatibility of (21) with the temporal adverbs at noon or at present and with the prepositional phrase by John should be quite obvious. The justification for the following claim is, however, not obvious at all:

> These two characteristics of the static passive, incompatibility with agent phrases and [incompatibility with definite Time Adverbs in the same clause], are also found in the periphrastic perfects under consideration here. (GALLAGHER 1970b: 54)

2.2. Without adducing any examples beyond those mentioned below, we would like to claim here that even a short look at a small set of sentences containing the Present Perfect will reveal that the English Perfect is completely compatible with the occurrence of an agent phrase. In addition to that, even Gallagher herself emphasizes that the agent mentioned in the subject-NP of Perfect sentences is required to be alive.[4] Therefore, her claim about the incompatibility with agent phrases cannot be taken to refer to the surface, but only to (part of) the underlying structure of the English (Resultative) Perfect, and our question must be: can the assumptions represented in tree-diagram (18b) be justified?

The agent-less sentence (20), which represents a possible surface structure of the underlying embedded S (BE) job finished in (18b) is in fact closely related to sentence (18a) by the relation of implication: the truth of (18a) implies the truth of (20), and one might consider it an advantage that the structure underlying the implied sentence ((20)) is part of the deep structure of the implying sentence ((18)). However, such an implicational relation does not only hold between a Present Perfect sentence (like (18a)) and (20), but also between a sentence in the Past (like (23)) and (20):

(23) John finished that job yesterday.

In addition to that, this latter implication is not due to the element Past, but to the fact that the verb (phrase) is "telic" (GAREY 1957) or - in Vendler's terms - an "accomplishment" (VENDLER 1957). As exactly the same situation obtains in respect of (18a) and (20), there is no reason[5] why an agent-less "static passive" structure should be assumed to occur in the deep structure of the Resultative Perfect, to which it is not related in any exclusive way at all.

2.3. As the above quotation (see 2.1.) indicates, some justification for an underlying "static passive" is sought in the historical development of English where sentences with have often had a word order as in sentence (19). It should be obvious that great methodological problems are involved in using diachronic arguments in order to support synchronic analyses. Without going into these problems, we would like to point out here that Gallagher's implicit assumption that the underlying structure diagrammed in (18b) will also underlie such earlier forms, might at first sight appear to find some support. Such support could be seen in the fact that inflectional endings indicated the close relationship between the Object and the Past Participle in Old English times. However, even though OE sentences frequently do not permit the addition of any agent-phrases (by-phrases) in surface structure, data from the history of English do not support Gallagher's assumption. As we can for example see from a sentence like (24),

(24) Ond hie ealle on þone cyning wæron feohtende, oþ þæt hie hine ofslægenne hæfdon. (OE Chronicle, about 755; quoted from VISSER 1973: 2387)

the referent of the subject-NP of have, which still shows traces of a possessive meaning in OE, must also be taken to be the "agent" of the Object-Participle-construction, i.e., we can assume the occurrence of two (lexically and referentially) identical noun-phrases in the deep structure of such OE sentences.[6] The fact that there is a whole set of examples (including more recent ones) in VISSER (1973: §2118), where the referent of the (surface) subject-NP does not at the same time denote the person that actually performed the action (as e.g. in (25))

(25) He hadde þare tweie castles bi-walled swiþe faste. (Layamon, about 1205; quoted from VISSER 1973: 2388)

supports the view that there are two underlying (active) sentences in deep structure, none of them without a subject-NP. This applies the more so, as all of Visser's examples permit the insertion of by-phrases, although only three of them actually contain one: the reason is that the respective contexts permit the deletion of the second agent-phrase in all other cases.

That the later development of English finally resulted in an increasing confusion of the two structures represented by (24) and (25) is therefore not due to an agentless deep structure involved at an earlier period, as Mary Gallagher claims:

> One result of the fact that the static passive does not indicate an agent is that the older form of the periphrastic perfect is ambiguous with respect to agent. (GALLAGHER 1970b: 54)

The ambiguity (and the subsequent development of the modern Present Perfect) could rather be said to result from the identical word order in surface structure, as Visser points out:

> After about Shakespeare's time the pattern with post-position of the object gradually became the normal one. One of the causes of the declining of the pattern with mid-position of the object may have been the fact that it became increasingly liable to be confused with a pattern with the same surface structure but an entirely different deep structure, viz. the [structure] of the type _he had it done_ = _he got it done_ (_by someone else_) ... (VISSER 1973: 2190)

If we also take into consideration that this change in word-order coincided (a) with the loss of the originally close relationship between the Past Participle and the Object and (b) with the loss of the 'possession'-meaning in _have_, we must say that it is disregard of these important historical changes which ultimately leads to Gallagher's assumption that there is a structurally relevant continuing relationship between the possessive and the perfective _have_, while her assumption of an agentless sentence in the deep structure of the modern English Present Perfect is not entirely unrelated to a misinterpretation of the Old English word order (Object - Participle).

In this connection, we should finally point out that the above-mentioned ambiguity "with respect to agent" (GALLAGHER 1970b: 54) which certainly could be observed at some stage in the history of English, cannot be observed any more, even though Gallagher claims that for sentences like (26):

(26) The landlord has thrown out the tenants. (GALLAGHER 1970b: 55)

Obviously, this sentence does not only mean that the landlord did it "with his own hands"; he may just as well have asked the police to do it for him. That (26) can be understood in these ways is due to the fact that our real world knowledge comes into play here, but it is certainly no justification for having an agentless static passive in the underlying structure: on the basis of our preceding discussion, we can safely abandon this idea. - If this were not so, one might wonder, whether a similar solution would not also be required in the case of a P a s t sentence like (27)

(27) Caesar built a bridge across the Rhine.

in order to make sure that the linguistic deep structure accounts for the fact that it certainly was not Caesar h i m s e l f who did the building.

2.4. Leaving our arguments against the "agentless" character of the underlying S _job finished_ in (18b) aside, we must still ask, whether there is any justification in postulating that this constituent should not admit any adverb. While it should be obvious that a sentence like (21) cannot take the adverbials marked by an asterisk, it should be just as obvious that a sentence like

(28) At present I've read only two of the books. (HUDDLESTON 1969: 784)

is perfectly acceptable and that (18b) must be able to account for it.

With a view to a semantic interpretation of (18b) the above explanation of this tree-diagram in terms of "result" (job finished) and "agent" (John) appears to be fully justified at first sight, if we keep its limitation to the Resultative Perfect in mind. Looking at (28) in a similar way, we might say that the "result" is books (are) read, while the "agent" is I; the adverbial at present here refers to the situation when the "result" was available, i.e. to the moment of speech. Given (18b), we cannot see any other reasonable assignment of the adverbial; however, this does not agree with the "adverbless" requirement imposed by Mary Gallagher on job finished or on books read, and it provides a further indication that (18b) cannot cope with actual language data and that it must be abandoned.

2.5. After we have employed Gallagher's term "static passive" for sentences like (21) and also (22b), we must finally ask, whether the terminology is entirely justified. This is necessary in view of the fact that the criterion "agentless" can also be applied to complement sentences with adjectives, as the substitutions in (29) and (30) indicate:

$$\text{(29)}\quad \text{The toast} \begin{Bmatrix} \text{is} \\ \text{seems} \end{Bmatrix} \begin{Bmatrix} \text{burnt} \\ \text{black} \\ \text{brown} \\ \text{crispy} \end{Bmatrix} \text{(*by Peter)}$$

$$\text{(30)}\quad \text{The job} \begin{Bmatrix} \text{is} \\ \text{seems} \end{Bmatrix} \begin{Bmatrix} \text{finished} \\ \text{hard} \\ \text{tiring} \\ \text{demanding} \end{Bmatrix} \text{(*by Paul)}$$

It is not only so that burnt and finished can be substituted by adjectives like black, hard, etc.; be can also be substituted by seem which - according to QUIRK et al. (1972: 241) - is "one of the diagnostic criteria for adjectives". Accordingly, these substitutions indicate, how problematic the "static passive" ultimately is: we can neither accept it as a part of the underlying structure of the Present Perfect, nor can the term itself be justified.

3. Instead of placing any further question-marks against Mary Gallagher's theory, I would now like to point out briefly, which facts must be considered in a proper analysis of the Present Perfect in English.

3.1. It is absolutely essential to distinguish three different elements: the time (or point) of speech, the point of reference, and the time at which the respective events take place (event-time). Whereas speech- and event-time are fairly straightforward concepts, this cannot be said about the time of reference. This concept is certainly not equivalent to what Mary Gallagher uses the term "reference" for:

> Certain Present Perfects ... do appear to make reference to the present time:
> John has been here since Saturday.
> Others, however, contain a Present ending, but appear to refer only to the past:
> John has thrown away his old books. (GALLAGHER 1970b: 72)

If the Perfect could really refer to present as well as to past time, it would be difficult to arrive at any unified description. Actually, the point of reference of the Present Perfect in English is the p r e s e n t ; it is identical with the speech time, whereas the event-time always precedes the speech- and reference-time. To represent this appropriately, it is best to assume a deep structure which essentially consists of two parts and therefore can accommodate speech-/reference-time (Present) and event-time (Past) at the same time. The adverbial at present (from sentence (28)), which provided the greatest problems for an analysis in terms of (18b), can easily be accommodated in such a system: as it does not refer to the time when the reading took place, it will have to appear under the same node as the speech-/reference-time. The resulting tree-diagram is analogous to one in HUDDLESTON (1969: 782):

(31)

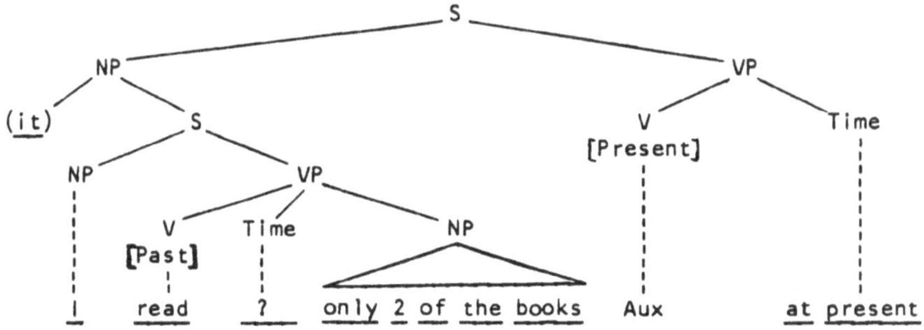

The differences between (18b) and (31) are obvious: firstly, the "result" is not represented any more, and therefore the diagram is not limited to the Resultative Perfect; secondly, what would have been the constituents of Mary Gallagher's "static passive" appears - t o g e t h e r with the "agent" - in an underlying "active" sentence. Only one problem-area remains, and this - incidentally - also provides the greatest problems for learners of English: the problem consists in the fact that the Perfect in English does not permit any past time adverb, or, in other words, it does not permit an explicit specification of the event-time. Even though diagram (31) might appear to be more elegant with the question-mark replaced by a past time adverbial, such a solution requiring the later (obligatory) deletion of the same adverbial clearly could not be justified here due to that characteristic limitation of the English Perfect. An acceptable approach might, however, be to explain the impossibility of an adverbial specification by attributing it to the fact that an underlying e m b e d d e d Past is involved. Such an explanation might be justified, if a similar limitation could be established for a past embedded in a past, i.e. for the P a s t Perfect.

3.2. Actually, the situation is not completely clear in the case of the Past Perfect, and Carlota S. SMITH (1976) does not only say that a sentence like

(32) Pursewarden had left Alexandria in November.

is ambiguous between an 'already'-reading (the event happened before the reference time which is given by in November) and a second reading, where he left i n N o v e m b e r , but she also adds:

> most speakers that I have asked find it odd and somehow incomplete. What is odd about [32] on the second reading is that it lacks a[n explicit] reference time. This is apparent from [33], in which [32] is embedded as a complement: the matrix sentence provides RT [the reference time]:
> [33] I remembered that Pursewarden had left Alexandria in November. (SMITH 1976: 5f.)

As this statement suggests a tree-diagram (for reading one) with an e m b e d d i n g Past specified by in November and an e m b e d d e d Past not specified at all, the situation might be parallel to that of the Present Perfect. Only when the embedding Past is "specified" by the Tense of a higher sentence (here: I remembered) does the embedded Past seem to permit a specification by a temporal adverb. Whether this statement applies to sentence (33) exclusively, or whether it has a more general application, must be further investigated.

3.3. Finally, let me just mention that a complete description of the Present Perfect in English must also consider the "character" of the verb or verb phrase. Here we can again refer to GAREY (1957) and his distinction between "telic" and "atelic" verbs which was applied to the English Perfect by Gero BAUER (1970). This distinction is important, as the traditional differentiation between different kinds of Perfect (Continuative, Iterative, Resultative etc.) can be attributed to the telic or atelic "character" of the lexical verb together with the use of the respective adverbs: the traditional differentiations do not have to be attributed to the Perfect as such. The Present Perfect itself has a unified meaning, for which it is essential that an action (or state) is referred to which happened "anterior to a certain point of reference [and which] is, at the point of reference, an accomplished fact" (BAUER 1970: 192). The point of reference may be specified adverbially, while the time of the event must not be specified, except for its beginning or for its duration as in (34):

(34) I have been reading for quite some time now.

NOTES

* Thanks are due to Miss J. Carter, P. H. Marsden, Ph.D., and Dr. W. Nöth for their comments on an earlier version of this paper.

1 Gallagher mentions (3b) without any further comments on its grammaticality or acceptability. As regards the other examples under (3) and (4), we added the "*?" in order to indicate that the sentences are fairly awkward (as under (3)) or - for two British informants - entirely unacceptable without -ing ((4c) and (4d)).

2 That a kind of "back-shifting" (Jespersen's term; cf. JESPERSEN 1931: 151 ff.) or "re-orientation" (cf. HUDDLESTON 1969: 794) is involved in sentences like (7a) is brought out by a sentence like (7b), which is perfectly acceptable:
(7b) The man who answered the telephone yesterday is Henry's brother.
Applying the same test to (8a) results in the unacceptable sentence (8b),
(8b) *The woman who answered the telephone is Henry's girlfriend last week.
which indicates that an underlying "original" past is involved in the embedding as well as in the embedded sentence. The latter conclusion is supported by the fully acceptable example (8c):
(8c) The woman who answered the phone five minutes ago was Henry's girlfriend last week.

3 A sentence like (10b)
(10b) Paul's father was tall when he was a boy.
is of course completely acceptable, as it does not claim that Paul's father has become smaller, but that he once was "tall for a boy", which - in the course of time - does not automatically lead to a state which might be described as "tall for an adult human male". - Similar considerations apply in the case of sentences like (10c):
(10c) Young Johnny is tall.

4 It should, however, be pointed out that - contrary to Mary Gallagher's claim - the "alive-subject"-constraint does not operate in sentences like (35a), in spite of the occurrence of have:
(35a) John is said to have done that work very well.
This becomes more obvious when we consider sentences where the subject-NP refers to a person who is known to be long dead, as in (35b) and (35c):
(35b) Shakespeare is said to have written impressive dramas.
(35c) Shakespeare is said to have been a drunkard.
(The underlying structures of sentences like (35a-c) are discussed more extensively by HOFMANN 1966.)

5 At least one of Mary Gallagher's reasons is that she sees a parallel between sentences like (20) (with be) and (18a) (with have) on the one hand

and sentences like (36a) and (36b) on the other:
(36a) A book i s on the table.
(36b) The table h a s a book on it.
Sentences (36a) and (36b) are derived from the same underlying structure ((BE) a book on the table); at the same time, Gallagher's transformational derivation of (36b) resembles closely her derivation of (18a) from (18b). However, while (36a) and (36b) are broadly synonymous, (18a) and (20) are only related implicationally (see above).

6 We assume that OE sentences like hie hine ofslaegenne haefdon are derived from the two underlying active sentences hie haefdon and hie hine ofslogon.

REFERENCES

BALD, Wolf-Dietrich (1972): Studien zu den kopulativen Verben des Englischen. Munich: Hueber.

BAUER, Gero (1970): "The English 'Perfect' reconsidered". Journal of Linguistics 6: 189-198.

CHOMSKY, Noam A. (1965): Aspects of the theory of syntax. Cambridge, Mass.: MIT Press.

GALLAGHER, Mary J. (1970a): "Adverbs of time and tense". Papers from the sixth regional meeting, Chicago Linguistic Society: 220-225.

--- (1970b): Have and the Perfect in English. Ann Arbor, Mich.: Univ. Microfilms (Film 70-13,322; Univ. of Illinois at Urbana-Champaign Ph.D. Diss.)

GAREY, Howard B. (1957): "Verbal aspect in French". Language 33: 91-110.

HOFMANN, T. Ronald (1966): "Past Tense replacement and the modal system". Report no. NSF-17 (Mathem. linguistics and automatic translation) to the Nat. Science Foundation. Cambridge, Mass.: Harvard U.: VII-1 to VII-21.

HUDDLESTON, Rodney D. (1969): "Some observations on tense and deixis in English". Language 45: 777-806.

JESPERSEN, Otto (1931): A modern English grammar on historical principles. Part IV: Syntax. London: Allen & Unwin. Copenhagen: Munksgaard.

QUIRK, Randolph / GREENBAUM, Sidney / LEECH, Geoffrey / SVARTVIK, Jan (1972): A grammar of contemporary English. London: Longman.

SCHOPF, Alfred (1963): "He is being clever". Anglia 81: 267-297.

--- (1969): Untersuchungen zur Wechselbeziehung zwischen Grammatik und Lexik im Englischen. Berlin: de Gruyter.

SMITH, Carlota S. (1976): A theory of auxiliary have in English. Bloomington, Ind.: Indiana University Linguistics Club (mimeographed).

VENDLER, Zeno (1957): "Verbs and times". The Philosophical Review 66: 143-160.

VISSER, F. Th. (1973): An historical syntax of the English language. Part III, second half: Syntactical units with two and with more verbs. Leiden: Brill.

THE SYNTAX OF GERUNDS IN ITALIAN*

Marina Nespor

1. Italian Gerunds always end in -ndo. Thus they are neither marked for person nor for number and gender. Examples of sentences containing gerunds are given in (1) and (2) below, where the so called Gerundio Presente and Gerundio Passato respectively are exemplified[1].

(1) Maria nuota pensando alle frasi relative.
'Mary swims (while) thinking about relative clauses.'

(2) Giorgio se ne andò, avendo finito di lavorare.
'George left, having finished to work.'

As can be seen from (3) and (4) below, the terms presente and passato do not refer to the time in which the sentence is uttered, as is the case with other tenses in Italian, but rather to the time to which the finite verb of the main sentence refers. Presente then means simultaneous and passato anterior[2].

(3) a. Giocando perse tutto.
'He lost everything in games.'
b. Sbagliando s'impara.
'One learns by making mistakes.'
c. Ridendo e scherzando arriverò fino in Turchia.
'Loughing and joking I will get to Turkey.'

(4) a. Avendo finito il pranzo uscì.
'Having finished his/her lunch he/she left.'
b. Avendo finito tutto esce e scappa.
'Having finished everything he/she gets out and runs away.'
c. Avendolo finito potrà andarsene.
'Having finished it he/she will be able to go away.'

The examples a, b and c of (3) and (4) above show that the finite verb can be past, present or future respectively.

Gerunds can also have different meanings, as can be seen in (5)-(9).

(5) Giorgio è scappato, avendo preso paura.
'George ran away since he got scared.'

(6) Sbagliando s'impara.
'One learns by making mistakes.'

(7) Raschiando vien via.
'It comes away if you scrape it.'

(8) Giorgio ha visto Maria attraversando la strada.
'George saw Mary while crossing the street.'

(9) Giorgio, credendo di farle piacere, le ha regalato un gelsomino.
'George, thinking to please her, gave her a jasmin.'

They vary also in that they occur in different positions in the sentence. That is initially, as in (6) and (7), finally, as in (8) and internally, as in (9). In addition, they can have different intonations. For instance they may or may not be separated from the rest of the sentence by a pause[3].

In this paper I am going to analyse the syntactic structure of gerunds in Italian[4]. I will not take into consideration the semantic differences mentioned above, since I believe that it is useful, for the time being, to keep syntax separate from semantics. For the sake of clarity I will include in my examples only sentence final gerunds and will show at the end that the fact that they do occur in other positions as well follows in a straightforward way from my analysis.

2.0. Assuming that differences in the phonetic pattern of speech might signal differences in the syntactic structure, I will treat separately gerunds with different intonation patterns.

2.1.1. Let us start by considering gerunds which are separated from the rest of the sentence by a pause. (Let's call them G1) I wish to argue that they are dominated by an S node, although the subject $\bar{\bar{N}}$ often fails to surface. The first argument is based on the fact that the subject $\bar{\bar{N}}$ can actually appear in the surface, as is seen in (10)-(12).

(10) Arriverò, Dio volendo.
'I will arrive, God willing.'

(11) Partiremo domani, il tempo permettendolo.
'We will leave tomorrow, weather allowing it.'

(12) Ci credo, avendolo detto Giorgio.
'I believe it, George having said it.'

Another argument in favour of the sentence-hood of gerunds can be made if we assume a theory (as for instance the standard theory) that considers Passivization to be clause internal. The argument then is simply that this transformation can apply within gerunds as is seen in (13), where Passivization has applied twice.

(13) Maria è stata consolata da Mario essendo stata trattata male da Giorgio.
'Mary has been consoled by Mario, having been mistreated by George.'

If, on the other hand, we do not consider Passivization to be clause internal, (13) is evidence that the gerund is dominated by a cyclic node (i.e. either S or NP). Since gerunds cannot be NP's we can conclude that they are sentences[5].

A third argument in favour of the sentence-hood of G1's is based on the fact that root transformations[6] can apply to them as is seen in (14) where in b Topicalization has applied.

(14) a. Paolo ha baciato Francesca, vedendo avvicinarsi la fine.
'Paul kissed Francesca, foreseeing the end.'
b. Paolo ha baciato Francesca, la fine vedendo avvicinarsi.

2.1.2. In the previous section I concluded that G1's are dominated by an S node and not simply by a V. I claim in addition that the gerundive S is a sister of the main clause, not a subordinate clause. As the first argument in favour of the sisterhood of gerunds I will show that root transformations, such as Topicalization or Parenthetical Formation can apply to both sentences. In (15) an example is given in which Parenthetical Formation has applied to both S nodes and in (16) an example is given in which Topicalization does the same.

(15) a. Paolo ha picchiato Francesca, credo, pensando ad altro, mi hanno detto.
'Paul beat Francesca, I think, (while he was) thinking about something else, I have been told.'
b. Paolo ha picchiato, credo, Francesca, pensando, mi hanno detto, ad altro.

(16) a. Paolo ha baciato Francesca vedendo avvicinarsi la fine.
b. Francesca Paolo ha baciato, la fine vedendo avvicinarsi.

Since root transformations apply only to root sentences, these examples show that G1's are in root sentences and are therefore sisters to the main clause. A sentence such as (17) will then have a structure like (18)

(17) Marco andrà in Grecia, il tempo permettendolo.
'Mark will go to Greece, weather allowing it.'

(18)

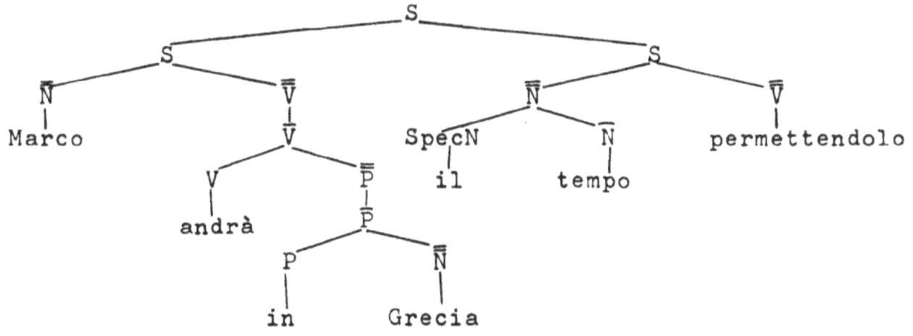

2.1.3. A further argument that confirms the analysis just offered is based on the intricate facts of reference of gerunds, which seem to be very similar to those of coordinate structures.

When the subject of the gerund surfaces as in (17) or when the subject is obvious, for instance because of a reflexive as in (19) or because of the semantics of the sentence, as in (20) (because one person cannot be passing by a house and going out of it at the same time), there are no problems as far as reference goes.

(19) Essendomi fatta male fu l'unico che mi soccorse.
'Having hurt myself, he was the only one who came to help me.'

(20) Ho visto Guendalina, essendo passata sotto casa proprio mentre uscivo.
'I've seen Guendalina, having passed by (my) house just when I was getting out.'

In all other cases the gerund can only refer to the subject of the other coordinate, and not to any of its complements, as is seen in (21) and (22).

(21) Ho visto Guendalina partendo da Sydney.
'I've seen Guendaline while leaving Sydney.' (While I was leaving Sydney./*while she was leaving Sydney.)

(22) Romeo ha litigato con Giulietta non potendo fare niente di meglio.
'Romeo fought with Giulietta, non being able to do anything better.' (he was not able to do anything better/*she was not able to do anything better)

If the arguments given above in favour of the sentencehood and sisterhood of G1's have been convincing, the interpretation of (21) and (22) should not be surprising since similar facts can be noticed in other coordinate constructions, as is seen in (23) and (24).

(23) Luca baciò Maria e se ne andò.
'Luca kissed Mary and left.' (he left/*she left)

(24) Luca comprò una rosa per Luigi e la buttò via.
'Luca bought a rose for Luigi and threw it away.'
(Luca threw it away/*Luigi threw it away)

From these facts it seems that there is a constraint on coordinate structures that says that when the subject of one of the conjuncts does not surface, the understood subject is the subject of the other conjunct and not any of its other NP's. As can be seen from (25), where Passivization has applied to the first conjunct, this is a surface constraint, since it is the surface subject (and not the deep subject) of the first conjunct that is understood as subject of the second conjunct

And as can be seen in (26) and (27) the same can be said for G1's, thus confirming the thesis that G1's are coordinated sentences.

(25) Maria fu picchiata da Marco e pianse.
'Mary was beaten by Mark and cried.' (she cried/*he cried)

(26) Maria fu vista da Marco, avendo attraversato il parco in bicicletta.
'Mary was seen by Mark, having crossed the park by bike.' (she had crossed/*he had crossed)

(27) Maria fu vista da Marco, avendo aperto la finestra.
'Mary was seen by Mark, having opened the window.' (she had opened the window/*he had opened the window)

2.2.1. Let us now consider gerunds that are not separated from the rest of the sentence by a pause. Let us call them G2's. In the present section I will argue that their structure is different from that of the gerunds considered in section 2.1. above. The arguments I give are based on the fact that none of the arguments given there in favour of the sentencehood of G1's holds for G2's. The first argument was based on the fact that there are sentences in which the subject of G1's surfaces. No such examples exist for G2's or at least I did not find any.

The second argument was based on the fact that Passivization can apply to G1. Also this is not the case for G2, as can be seen in (28) where a contains a G2 followed by a direct object, and b is its counterpart in which Passivization has applied.

(28) a. Maria parla mangiando la torta.
'Mary speaks eating the cake.'
b.*Maria parla la torta essendo mangiata da lei.

The third argument in favour of the sentencehood of G1's was based on the fact that a root transformation can apply to them. But this, too, is not true for G2's, as is seen in (29).

(29) a. Paolo cantava mangiando la torta.
'Paul was singing (while) eating the cake.'
b.*Paolo cantava la torta mangiando.

From these three arguments I conclude that G2's are not sentences but simple $\bar{\bar{V}}$'s. The conclusion that G1 and G2 have different syntactic structures is confirmed by the fact that there are sentences that can be read both with and without a pause preceding the gerund and that have two different meanings for the two intonations. An example is given in (30).

(30) Marco fa ginnastica andando in bicicletta.
'Mark exercises while going by bike.' or:
'Mark exercises by going by bike.'

2.2.2. We can now ask ourselves what the structure of G2's is. I am going to argue that they are complements of the verb, probably dominated by a node $\bar{\bar{A}}$[10].

The first argument is based on the fact that the verb alone cannot be extracted from a sentence such as (31a), as can be seen from the ungrammaticality of (31b).

(31) a. Maria voleva mangiare cantando.
'Mary wanted to eat while singing.'
b.*Era mangiare che Maria voleva cantando.
'It was eating that Mary wanted while singing.'

Instead the verb plus the gerund can be extracted, as is seen in (32).

(32) Era mangiare cantando che Maria voleva.
'It was eating while singing that Mary wanted.'

If we assume that only constituents may move, then we can conclude that G2 forms one constituent with V and is therefore its complement.

The second argument in favour of considering G2 as a complement of V is that it can stand between V and other complements of V, as is seen in (33).

(33) Marco corre strillando verso casa.
'Mark runs screaming towards home.'

The only element that can stand between V and a complement of V, in addition to another complement, is a parenthetical. But in (33) <u>strillando</u> does not have the intonation that parentheticals have. Therefore it is a complement of V. The structure of a sentence such as (31a) is then (34).

(34)

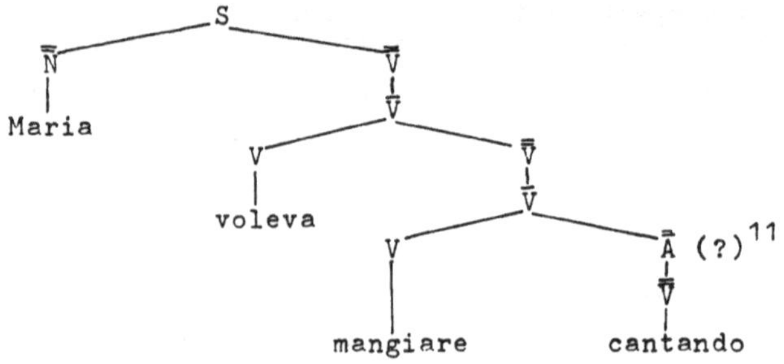

3. S-initial and S-internal Gerunds.

3.0. So far we have considered only gerunds in S-final position. As I said at the beginning, however, gerunds can also appear S-initially and S-internally. S-initial gerunds can be either followed by a pause or not and S-internal gerunds are always both preceded and followed by a pause.

3.1. Let us consider first G1's. Since they are coordinated to the clause with the finite verb, they can either precede it or follow it. And since when they occur S-internally they have the same intonation as parentheticals[12], the reason they can appear there is obvious. An example of a G1 in final and initial position is given in (35) a, b respectively and examples of G1 in internal position are given in (35) c and d.

(35) a. Giorgio le ha regalato un gelsomino, credendo di farle piacere.
b. Credendo di farle piacere, Giorgio le ha regalato un gelsomino.
c. Giorgio, credendo di farle piacere, le ha regalato un gelsomino.
d. Giorgio le ha regalato, credendo di farle piacere, un gelsomino.

3.2. In 2.2.2. I claimed that G2's are complements of the verb. This claim is confirmed by the fact that they can occur pre-verbally (S-internal), post-verbally either before another complement of V (S-internal) or after all of its complements (V-final and perhaps also S-final). Examples are given in (36) a, b and c respectively.

(36) a. Giorgio ridendo corre a Roma.
'George laughing runs to Rome.'
b. Giorgio corre ridendo a Roma.
c. Giorgio corre a Roma ridendo.

Moreover, as other complements, G2's can be topicalized and thus occur S-initially, as seen in (37).

(37) Ridendo Giorgio corre a Roma.

4. Conclusions.

To summarize briefly, I have distinguished two possible structures in which non-periphrastical gerunds may occur: 1) in coordinate structures 2) in the complement of V. With this analysis I have accounted for both the syntactic behaviour of gerunds and their reference[13].

FOOTNOTES

Some of the ideas of this paper are taken from section 3.7 of my doctoral dissertation. I would like to thank the people who read and commented on that section, in particular Donna Jo Napoli and Robert Rodman. In addition I thank Ivonne Bordelois for a number of suggestions and Enzo Lo Cascio and Donna Jo Napoli who carefully read the first draft of this paper, for their many helpful insights. A final thanks goes to my informants for their patience.

1. See, among others, Regula M. and J. Jerney (1965), Pittano G. (1972) and Altieri Biagi M.L. and L. Heilman (1973).

2. See Pittano G. (1972).

3. The presence or absence of a pause is the most obvious difference between the two intonation patterns. However, I will not discuss these patterns in detail here, since, for the purpose of the present discussion, it is sufficient to know that they are different even in fast speech where pauses might be absent.

4. Gerunds can also appear in the so-called periphrastical constructions, following the verbs stare, andare and, more seldom, venire. Examples are given in (i) - (iii).
   (i) Sto suonando il flauto. 'I am playing the flute.'
   (ii) Va dicendo storie a destra e a manca. 'He goes around telling stories.'
   (iii) Ma cosa vien raccontando? 'But what is he saying?'
   In this construction stare, andare and venire behave very much like auxiliaries, but for reasons of space I will not analyse this in detail here.

5. I owe this observation to D.J. Napoli.

6. For the notion of Root Transformation and Root Sentence see Emonds (1976).

7. While most speakers accept (14b) only in a very formal style, I found some who did not accept it at all. Therefore this argument holds only for some varieties of Italian. For more examples of this kind see Nespor (1977: 90-92).

8. For (15b) see ftn. 7 above.

9. These sentences have been chosen in such a way that, semantically, the gerund could refer to both the subject and the complement.

10. In Fornaciari (1881) it is observed that some gerunds behave as adverbials.

11. Notice that there is a question mark near $\bar{A}$ in (34). This is because I do not know exactly what the internal structure of that node is, or even whether it is an $\bar{A}$ at all. The only thing I showed here is that there is no S node.

12. See Emonds (1976: 43-60) and Nespor (1977: 79-87).

13. If this analysis is correct, a sentence such as (i)
   (i) Luca ha visto Giorgio mangiare attraversando il parco. 'Luca saw George eating while crossing the park.' (while Luca was crossing the park/while George was crossing the park)

where both Luca and Giorgio can be understood as subjects of the gerund, supports an analysis for sentences with infinitivals embedded under perception predicates which does not make use of a rule of subject to object raising. In this analysis Giorgio is the subject of mangiare during the whole derivation. (For arguments in favour of this analysis see Nespor (1977) and the works cited there.) That is, if either Luca or Giorgio can be understood as subjects of the gerund, according to what I said in 2.1.3. above, both Luca and Giorgio should be the surface subjects of some sentence.

BIBLIOGRAPHY

ALTIERI BIAGI, Maria Luisa/HEILMAN, Luigi (1973): La lingua italiana. Mursia.

CHOMSKY, Noam (1973): "Conditions on Transformations". S. ANDERSON/ P. KIPARSKY (eds.): A festschrift for Morris Halle. Holt, Rinehart and Winston Inc.

EMONDS, Joseph (1976): A Transformational Approach to English Syntax. Academic Press Inc.

FORNACIARI, Raffaello (1881): Sintassi Italiana. Ristampa (1974). Biblioteca Sansoni.

NESPOR, Marina (1977): Some syntactic structures of Italian and their relationship to the Phenomenon of Raddoppiamento Sintattico. Unpublished doctoral dissertation. U.N.C. North Carolina.

PITTANO, Giuseppe (1972): Grammatica Italiana. Mondadori.

RADFORD, Andrew (1974): Bidirectionality in Raising. Unpublished doctoral dissertation. Trinity College, Cambridge.

REGULA, M./JERNEY, J. (1965): Grammatica Italiana Descrittiva. Francke Verlag, Bern und München.

SCHACHTER, Paul (1977): "Constraints in coordination". Language 53.1.

ON ITALIAN CLEFT SENTENCES*

Giorgio Graffi

In this paper, I will deal with some problems arising from an analysis of Italian cleft sentences; more precisely, I will be concerned with the cases where the "focussed" phrase is a personal pronoun, for instance:

(1) a. Sono io che l'ho detto
"It's I that said it"
b. Sei tu che l'hai detto
"It's you that said it"
c. E' lui che l'ha detto
"It's he that said it"
d. Siamo noi che l'abbiamo detto
"It's we that said it"
e. Siete voi che l'avete detto
"It's you that said it"
f. Sono loro che l'hanno detto
"It's they that said it"

I shall try to account for the following facts: (i) we must explain the facts of person and number concord between the verb _essere_ (to be) and the focussed personal pronoun (this problem doesn't appear in languages like English or French, where the matrix verb is always of the form _C'est_ or _It's_, no matter what the following person be: _C'est moi_, _C'est vous_ or _It's I_, _It's you_); (ii) the second fact we have to account for is shown by the following examples:

(2) E' lui che Maria dice che è malato
"It's him that Mary says that he is ill"

(3) *Sei tu che Maria dice che è malato

"It's you that Mary says that he is ill"

(4) *Sono io che Maria dice che è malato

"It's me that Mary says that he is ill"

That is to say, we must explain why, in sentences such as (2) the pronoun seems to be able to 'rise' from a deeper position (in this case two cycles deeper), while in sentences like (3) and (4) this seems not to be the case. Note that the sentences are also ungrammatical with the most deeply embedded verb in the same person as the focussed pronoun[1]:

(5) *Sei tu che Maria dice che sei malato

"It's you that Mary says that you are ill"

(6) *Sono io che Maria dice che sono malato

"It's me that Mary says that I am ill"

Incidentally, sentences like (2) seem to invalidate the "Subjacency Condition" of CHOMSKY 1973, as BACH/HORN 1976 have already noted for a similar example:

(7) It was Harriet that Bill wanted John to kiss

(7) would be derived from a structure like

(8) It BE $\Delta[_S$ COMP Bill want $[_S$ COMP John to kiss Harriet]]

In order to derive (8), as Bach and Horn say, we would have to violate the Subjacency Condition, since substitution of Harriet for Δ would reach across two cyclic nodes.

PINKHAM/HANKAMER 1975, in a very interesting paper on cleft sentences, propose to derive them from two different sources, and precisely they say that cleft sentences which can also be interpreted as headless relatives, as for instance

(9) It was a banana that she was eating

are generated basically, while sentences such as

(10) It was during the war that he acquired his wounds

are generated by means of a rule of It-clefting which extracts a focus constituent from the clause and creates the new structure with It-BE. In other words, in structures like (10), the It-BE phrase is introduced transformationally. This "structure-building" rule is an unwelcome step, because it noticeably increases the power of transformations; and Pinkham and Hankamer explicitly stress the "challenging" character of their argument. In fact, it seems to have some empirical confirmation, for instance in the case of the following German sentences:

(11) Es ist Hans, der mit dir sprechen will

(12) Es ist neben mir, dass er sich hingesetzt hat

These sentences may prove a puzzle, since they both are (or both can be) cleft sentences, and yet they have a somewhat different form: the embedded clause in (11) is introduced by the relative pronoun, while the one in (12) is introduced by the complementizer dass. The Pinkham and Hankamer's analysis, which postulates two different origins for cleft sentences, would in my opinion account for these facts.

Now, turning back to our Italian examples, let's see how they would be treated in the Pinkham-Hankamer framework. First, let's consider the problem of agreement between the focussed pronoun and the verb BE, as in (1b):

(1) b. Sei tu che l'hai detto
"It's you that said it"

If we assume, with Pinkham and Hankamer, that some cleft sentences are transformationally derived by a structure-building rule which creates a structure of the type It-BE, then it is a quite obvious step to say that this rule also yields, in Italian, the verb-pronoun concord in person and number.

Secondly, in order to explain the ungrammaticality of structures like

(3) *Sei tu che Maria dice che è malato
"It's you that Mary says that he is ill"

we can assume that they behave in a similar way to sentences like

(13) Giuseppe dice che i Bianchi abitano a Pavia da 4 anni
"G. says that the B.'s have been living in P. for 4 years"

which cannot be transformed by cleft sentence formation into sentences like

(14) E' da 4 anni che Giuseppe dice che i Bianchi abitano a Pavia
"It's four years that G. says that the B.'s have been living in P."

(13) and (14) are both grammatical, but they are clearly not synonymous. If we assume, with Pinkham and Hankamer, that cleft sentence formation for PP derives from a structure-building transformation and not from a rule similar to relative clause formation, then we can explain the behavior of (13) and (14) by relating it to the behavior of (15) and (16), both grammatical but clearly not synonymous:

(15)=(13) Giuseppe dice che i Bianchi abitano a Pavia da 4 anni
"G. says that the B.'s have been living in P. for 4 years"

(16) Da quattro anni Giuseppe dice che i Bianchi abitano a Pavia
"It's four years that G. says that the B.'s have been living in P."

But we now see the weakness of Pinkham and Hankamer's proposal: first of all, as far as the movement of PP or of first and second person pronouns is concerned, the problem is simply shifted but not solved; I mean that, as in the case of (15) and (16), we are reminded that we are dealing with the same problems also in the case of non-cleft constructions. But this is not my main point; and I would add, as far as this goes, that

I am developing Pinkham and Hankamer's insights rather than quoting them. The biggest problem with Pinkham and Hankamer's proposal is that they make a neat distinction between cleft sentences with NP-focus, which can also be interpreted as headless relatives and which are generated basically, and cleft sentences with a non-nominal focus, which are generated by a structure building transformation. Indeed, we have to assign a NP status to third person pronouns and a non-NP status to first and second person pronouns, in order to explain the agreement facts of (1) and the difference between (2), on the one hand, and (3)-(6) on the other.

CHOMSKY 1976b has presented an approach to cleft sentences which can dispense us from having recourse to structure-building transformations; Chomsky proposes to generate basically the structure

(17) It - is - $\bar{\bar{S}}$

where $\bar{\bar{S}}$ indicates a sentence with its complementizer (in the sense of Bresnan (1970;1972)) and its TOP position; the base rules introduced by Chomsky to generate these structures are the following:

(18a) $\bar{\bar{S}} \rightarrow$ TOP $\bar{S}$

(18b) $\bar{S} \rightarrow$ COMP$\left\{ \begin{matrix} \bar{\bar{S}} \\ S \end{matrix} \right.$

Rule (18b) allows recursion for focussed structures; we get, in a structure like (17), the focussed phrase in the TOP position of $\bar{\bar{S}}$. This would also explain, in Chomsky's view, the obvious similarity in meaning between (19) and (20):

(19) The book, I read

(20) It is the book that I read

In the case of (19), for example, the focussed phrase would be generated basically in the TOP position and the deep structure of the sentence would look something like

(21) $[_{\bar{\bar{S}}}[_{TOP}\text{ the book}][_{\bar{S}}[_{COMP}\,\Delta][_{S}\text{ I read the }\underset{+WH}{\text{book}}]]]$
Then to this structure would apply the transformation of WH-movement in the COMP position, yielding

(22) $[_{\bar{\bar{S}}}[_{TOP}\text{the book}][_{\bar{S}}[_{COMP}\text{the book}][_{S}\text{I read }\underline{t}]]]$
The second occurrence of <u>the book</u> is deleted under identity with the phrase generated in the TOP position and so we reach the superficial form of the sentence:

(23)=(19) The book, I read

(20) would be analysed in an analogous way; the only difference would be that the structure $\bar{\bar{S}}$ is embedded in the <u>It-BE</u> structure, as rule (18b) allows.

Chomsky observes that sentences like (23) (the example is Pinkham and Hankamer's) cannot be treated as the ones above:

(23) It was only reluctantly that he agreed to swim at all

and he postulates, to explain them, that they are derived not by a rule of WH-movement, but by a rule of adverb-preposing into TOP position. He concludes: "We have in effect two sources for clefts but no separate rules; furthermore, we need not postulate 'structure-building rules'". In my opinion, it is surely positive that no structure-building rules are postulated. However, I think that, in order to explain the different behavior of (2), on the one hand, and of (3)-(6) on the other, Chomsky's analysis says no more than Pinkham and Hankamer's. Indeed, we cannot say that first and second person pronouns behave like adverbs, while third person pronouns behave like NP's (or, if we prefer, we cannot choose for <u>io</u> and <u>tu</u> an adverb-preposing analysis that we don't allow for <u>lui</u>).

Chomsky's proposal (i.e. the basic generation of the focussed phrase in the TOP position) seems unsatisfactory in another respect. Consider the following sentences:

(24) E' te che Maria dice che vuole sposare

"It's you that Mary says that she wants to marry"

(25) *Sei tu che Maria dice che vuole sposare
"It's you that Mary says that she wants to marry"
(26) *Sei tu che questa situazione danneggia
"It's you that this situation damages"
(27) Sei tu che sei danneggiato da questa situazione
"It's you that are damaged by this situation"

These sentences seem to suggest the traditional analysis in terms of movement (promotion, upgrading) of the personal pronoun from the embedded clause to the matrix: the pronoun has the feature [+ACC] in the embedded clause at the deep structure level, and it carries this feature with it as far as the surface position in the matrix sentence. The verb essere (to be) agrees with the focussed pronoun in person and number only if the pronoun is marked [+NOM]. If the personal pronoun were base-generated in the TOP position (according to Chomsky), we would be compelled to say that: 1) verb-pronoun agreement is optional (to explain the grammaticality of both (24) and (27)); 2) we have no recoverability of deletion if the focussed pronoun and the WH-element moved in the COMP position have different case-markers (to explain the ungrammaticality of (25) and (26)). But, while the first proposal sounds at least strange, the second is falsified from some examples of non-restrictive relatives, as L. Rizzi has pointed out to me:

(28) Anche tu, che credevamo un amico, ci hai traditi
"You too, whom we believed a friend, have betrayed us"
(29) Hanno licenziato anche te, che eri sempre stato tranquillo
"They have dismissed you too, who had always been quiet"

On the basis of these observations, I would propose for Italian cleft sentences a deep structure of this kind (similar to that proposed by RUWET 1974 for the French ones):

(30) Essere $\Delta[_{\bar{S}}$ che $[_S \ldots]]$

As we have already noted, the focussed element is moved to the position noted here by Δ; if this element is a personal pronoun and its case marker is [+NOM], the verb essere agrees with it in person and number.

Now let's turn back to the second group of questions we posed at the beginning of this paper. We see that the difference in grammatical status between (2), on the one hand, and (3) and (4) on the other can be easily captured by some rules of semantic interpretation: (3) and (4) are simply not interpretable, because no coreference to the personal pronoun in the matrix sentence can be assigned in the embedded clauses. This becomes clear, I think, from the English translation of both sentences. On the contrary, (2) is interpretable because the third person pronoun in the most deeply embedded clause, whose phonetic realization is zero, can be interpreted as coreferential to the personal pronoun in the matrix sentence.

Anyway, the proposal for the generation of Italian cleft sentences we have sketched above doesn't solve all our difficulties. I shall limit myself to elucidating them, because no solution I have tried to find seems to me fully satisfactory. The biggest difficulty arises again from sentences like (5) or (6), which I feel ungrammatical, although they cannot be ruled out by the rules of cyclic assignment of the case marker to the pronoun, which remains [+NOM], as it was in the deep structure, nor by interpretative rules, since the personal pronoun in the matrix sentence is surely to be interpreted as coreferential to the (unexpressed) second person pronoun in the most deeply embedded clause.

(5) *Sei tu che Maria dice che sei malato

(6) *Sono io che Maria dice che sono malato

An answer to these difficulties could be given by having recourse to the Chomskyan "conditions on transformations". We can say that (5) and (6) are out because of the Subjacency Condition: two coreferential pronouns are separated by two cyclic nodes. But CHOMSKY 1976a says that interpretative rules don't meet the Subjacency Condition; and why is this condition met in (5) and (6) but not in (2)?

We might observe, in an intuitive way, that, in sentences such as (5) or (6), there is a sort of "excess of information" which makes these sentences somewhat "odd": the indication of person and number of the focussed pronoun occurs twice, in the matrix clause and in the most deeply embedded one. (5) and (6) would be rejected, in this case, not on "grammatical" but on "perceptual" grounds. An argument in favor of this thesis might be the fact that, if in (5) or in (6) we replace the conjugated form with the infinitive in the most deeply embedded clause, the degree of grammaticality noticeably increases:

(31) Sei tu che Maria dice essere malato
"It's you that Mary says to be ill"

I acknowledge that (31) belongs to a highly formal style, but it seems to me that it is surely better than (5). The reason for this might be that the "excess of information" has disappeared.

But this explanation is clearly unsatisfactory: in effect, sentences such as the following, (32) and (33), are both perfectly grammatical (and the last one sounds much more formal and literary):

(32) Sei tu che sei malato
"It's you who are ill"

(33) Sei tu ad essere malato
"It's you who are ill"

In (32), we get the same "excess of information" as in (5); but (32) is completely acceptable. Our "perceptual" explanation is, therefore, inadequate, and we haven't found any good 'grammatical' solution to the problem: every proposal we have examined (Pinkham and Hankamer's, Chomsky's, and our own) fails to give a full solution to the second question we posed at the beginning of this paper. I hope that future works will give a more satisfactory treatment of this ticklish problem.

NOTES

* I am grateful to G. Cinque, M. Nespor and L. Rizzi, who have read an earlier draft of this paper and have offered me a number of suggestions. I also wish to thank J. Trumper. All errors are, of course, my own.

1 Not every native speaker of Italian feels as ungrammatical sentences such as (5) or (6). G. Cinque points out to me that two variants seem to exist, in a rather systematic way: in the first one, the focussed pronoun would still bring out the "function" which it has in the deep configuration ("accusative" vs. "nominative"); in the second variant there would be a general phenomenon of agreement between the personal pronoun and the verb *essere*. So, for instance, to express the meaning of the sentence *It's you that Mary wants*, we would have two different sentences, according to the two variants: speakers of the first variant would say *E' te che Maria vuole* (pronoun in the "accusative" case, no agreement between the pronoun and the verb *essere*), whereas speakers of the second variant would say *Sei tu che Maria vuole* (pronoun in the "nominative" case, concord between it and the verb *essere*). Speakers of the second variant would, of course, find grammatical (5) and (6), because their dialect has a rule of obligatory agreement between pronoun and verb. I must add that (5) and (6) are grammatical, or at least not fully out, also for some speakers of the first of the two variants.

BIBLIOGRAPHY

BACH, Emmon / HORN, George M. (1976): "Remarks on 'Conditions on transformations'". Linguistic Inquiry 7: 265-299.

BRESNAN, Joan (1970): "On complementizers: towards a syntactic theory of complement types". Foundations of Language 6: 297-321.

—— (1972): The theory of complementation in english syntax. Unpublished Doctoral dissertation, MIT, Cambridge, Mass.

CHOMSKY, Noam (1973): "Conditions on transformations". ANDERSON, Stephen R. / KIPARSKY, Paul (eds.) (1973): A Festschrift for Morris Halle. New York: Holt, Rinehart and Winston: 232-286.

—— (1976a): "Conditions on rules of grammar". Linguistic Analysis 2: 303-351.

—— (1976b): "On WH-movement". Unpublished manuscript.

PINKHAM, Jessie / HANKAMER, Jorge (1975): "Deep and shallow clefts". GROSSMAN, Robin E. / SAN, L. James / VANCE, Timothy J. (eds.) (1975): Papers from the eleventh regional meeting of the Chicago Linguistic Society. Chicago: Chicago Linguistic Society: 429-450.

RUWET, Nicolas (1974): "Les phrases copulatives en français". Recherches linguistiques 3: 143-191.

SUR L'IMPERSONNEL EN ITALIEN

Piero Caracciolo

L'exposé que vous allez entendre vous paraîtra peut-être anachronique. Anachronique en cela que l'appareil métalinguistique formalisant y joue un rôle moindre et pour ainsi dire d'appui, tandis que les faits dont nous allons traiter seront **plutôt** l'objet de réflexions grammaticales et sémantiques qui ne tendent pas à la systématisation.

Cela tient à une conception précise du rôle de la formalisation dans le cadre d'études portant sur une langue et non sur une théorie du langage. Dans ce cadre, pensons-nous, l'outil formel ne peut intervenir - au moins dans une première approche - qu'en tant que moyen de préciser certains concepts - tels les catégories grammaticales - autour desquels la description peut trouver des éléments d'ordre et de repère parmi les phénomènes observés.

Les phénomènes qui font l'objet du présent exposé sont les constructions italiennes utilisées pour exprimer l'impersonnel. On comprend dans l'impersonnel des phénomènes de différents ordres. D'une part - suivant une définition sémantique - des relations prédicatives dans lesquelles l'agent est indéterminé. Cela caractérise les énoncés passifs sans agent, comme *(1) Giulio è stato ferito* ou *(2) Giulio è rispettato.*

D'autre part, on regroupe sous l'impersonnel des constructions syntaxiques où le sujet est un pronom indéterminé comme dans *(3) uno queste cose non le fa;* et, enfin, les constructions caractérisées par le *si* dit impersonnel *(4) si lavora,* ou par le *si* dit "passivante": *(5) si lavora il vetro.* Il faut aussi rappeler les constructions comme: *piove, nevica,* que nous ne prendrons pas en examen.

Tous ces phénomènes méritent d'être rassemblés sous une même rubrique: le trait qui les unifie est l'indétermination de ce qui, dans la théorie à laquelle nous nous référons, est appelé la source de la relation primitive sousjacente à l'énoncé. Mais les problèmes qu'ils posent sont de type différent : pour le passif sans agent, il s'agit de voir quels sont les effets de l'effacement de l'agent avec les différents prédicats; avec le *si* "passivante" nous avons surtout des problèmes relatifs à l'ordre des mots, reliés aux propriétés primitives des prédicats et des arguments qui entrent en jeu.

Afin de traiter ces différents problèmes, nous utiliserons certains concepts de la théorie de l'énonciation d'Antoine Culioli, et notamment les concepts de relation prédicative, d'orientation et de terme de départ de la relation (1).

Du point de vue sémantique, l'impersonnel se présente comme une relation prédicative dans laquelle la source est tout à fait indéterminée. Nous pouvons presque établir une échelle d'indétermination: *Giovanni ha ferito Giulio; un soldato ha ferito Giulio; qualcuno ha ferito Giulio;* et ensuite l'impersonnel: *uno ha ferito Giulio, Giulio è stato ferito, Giulio si è ferito* (ce dernier pouvant être aussi réfléchi).

L'indétermination de la source paraît compatible avec toutes sortes de configuration des relations prédicatives. Si nous voulions adopter la terminologie de Benveniste (2), qui définit l'actif comme ces énoncés où "les verbes dénotent un procès qui s'accomplit à partir du sujet et hors de lui", alorsque "dans le moyen,..., le verbe indique un procès dont le sujet est le siège" (3), nous pourrions parler d'impersonnel compatible avec l'actif et le moyen. Le premier cas est celui du pronom *uno*, qui se présente comme un terme capable d'exprimer la source tout à fait indéterminée, dans des constructions où la source est en même temps le terme de départ. Ce type de construction est beaucoup plus fréquent en français (*on*), en anglais (*one*) et en allemand (*man*). (L'italien possédait autrefois un morphème apparenté

---

(1) CULIOLI, Antoine (1971): "A propos d'opérations intervenant dans le traitement formel des langues naturelles" in Mathématiques et Sciences Humaines, n°34 Paris. (voir,aussi,"note" à la fin de cette article).

(2) BENVENISTE, Emile (1950): "Actif et moyen dans le verbe" in Problèmes de Linguistique Générale, Paris 1966 Gallimard.

(3) Nous pensons que cette définition peut être reformulée en introduisant la notion de terme départ à la place de la notion de sujet.

au français *on* et à l'allemand *man*, le morphème *homo* ou *omo*, capable d'assurer ce type de fonction).

A côté de l'impersonnel actif, on trouve en italien, beaucoup plus fréquemment, l'impersonnel moyen, c'est-à-dire passif ou réfléchi.

Dans le passif sans agent, le terme de départ ne coïncide pas avec la source. La source est tout simplement effacée, et cela semble être possible seulement parce qu'elle n'est pas en même temps terme de départ (ainsi, le terme de départ semble devoir être réalisé en surface même lorsqu'il est indéterminé). Mais, l'éffacement de la source entraîne, selon les énoncés, des interprétations qui varient en fonction des propriétés primitives, des traits sémantiques des prédicats. Il nous faudra distinguer au moins trois classes de prédicats: ceux qui ont le comportement de *seguire*, ceux qui se construisent comme *rispettare* et ceux qui suivent *rompere*.

(6) Giulio è seguito et *(7) Giulio viene seguito* sont deux énoncés équivalents et ils peuvent être paraphrasés par:

Giulio viene seguito da chi lo segue

Cela peut être représenté par le fait que l'on construit un suiveur indéterminé et qu'on le prend en considération.

Par contre, une glose possible de:

(8) Giulio viene rispettato

(2) Giulio è rispettato

est:

Giulio viene rispettato da tutti

Ce type d'effacement de l'agent peut se représenter par l'introduction d'une opération de parcours sur toutes les sources possibles, de façon telle que la source serait non pas un individu indéterminé, mais la classe de tout ce qui peut être source dans la relation prédicative sousjacente à l'énoncé.

Dans *(9) il telefono è rotto*,

nous avons une suppression pure et simple de la source, une transformation de la relation *telefono-rompere-rompitore* en une relation entre *telefono* et *rompere*, liée à une valeur aspectuelle (étant résultant), analogue à celle que l'on retrouve dans les énoncés où le verbe *essere* n'est pas auxiliaire mais copule. *(10) Il telefono viene rotto*, par contre, peut se représenter de façon analogue à *(6)* et *(7)*.

L'étude du *si* dit "passivante" porte sur l'ordre des mots. Ce qui apparaît dans cette étude, c'est l'importance des propriétés primitives des prédicats, considérées non seulement pour chaque prédicat, mais pour des combinaisons de prédicats et notions. Quelques exemples suffiront à indiquer cette complexité.

(5) si lavora il vetro et

(11) il vetro si lavora

sont interprétés comme de l'impersonnel, quoique *(11)* se prête aussi à une interprétation modale de type devoir.

(12) si guarda la regina et

(13) la regina si guarda

présentent, par contre, des différences importantes. Le premier est sans doute un énoncé impersonnel, tandis que le deuxième peut être impersonnel (avec une modalité telle que: *la regina deve essere guardata*), ou bien réfléchi. Cela paraît tenir à la présence de la propriété ANIMÉ dans la notion REGINA et nous amène à formuler notre conclusion: la forme du réfléchi peut être employée pour exprimer l'impersonnel:

- soit avec un aménagement dans l'ordre des mots tel que la relation entre le but et la relation avec une place vide est reprise par un pronom (c'est le cas du *si* en début de phrase) qui exprime l'auto-repérage de cette relation, ce qui exclue alors la possibilité d'introduire un terme de départ;
- soit lorsque les propriétés primitives des termes impliqués sont telles qu'une relation réfléchie est impossible.

Alors, le pronom réfléchi (l'identification avec la source étant impossible) devient l'image de la relation elle-même, tandis que le terme instancié, à cause des propriétés primitives, ne peut assumer que le rôle de but.

Ceci peut être plus clairement exposé à l'aide d'une représentation qui contienne des éléments, pour ainsi dire, proto-formels: i.e. qui correspondent à certains éléments du métalangage de la Théorie de l'Enonciation qui sont définis par des propriétés formelles (voir "note" en fin d'article).

Nous indiquerons par R la relation de repérage. Etant donnée la relation:

(a) ( ) *lavora il vetro*

les propriétés primitives de *vetro* et *lavorare* font que la place vide de la source ne peut pas être instanciée par *vetro*; cela exclut que les constructions avec *si* puissent être interprêtées comme exprimant une relation réfléchie. La relation (a) doit être répérée par rapport à un terme de départ. Ce terme peut être la source indéterminée (si les propriétés primitives et les conditions modales le permettent). Cela donne:

(14) uno lavora il vetro.

On peut aussi repérer la relation (a) par rapport à son but, ce qui donne le passif sans agent:

(15) il vetro viene lavorato.

Il reste une troisième possibilité: l'auto-repérage, le repérage de la relation par rapport à elle-même. Le pronom *si* est la trace de ce type de repérage, que l'on peut représenter ainsi:

=

(b) ( ) *lavora il vetro* R *( ) lavora ( )*

=

ce qui peut être glosé par:
qualunque siano le condizioni e i soggetti: lavora il vetro chi lavora il vetro e chi lavora il vetro lavora il vetro.

Les propriétés primitives font que l'ordre des mots, en surface, ne change rien à l'interprétation. Nous avons aussi bien *(5)* que *(11)* (1). L'ordre des mots devient par contre pertinent avec les termes *regina* et *guardare* (cf. *(12)*, *(13)*), puisque *regina* peut être la source de la relation primitive et donc, la place non instanciée de cette relation peut être identifiée à *regina*. C'est là que l'observation intervient, et nous devons simplement remarquer que ce dernier type de relation est exprimé par:
la regina si guarda (avec modalité assertive),
tandis qu'une relation semblable à (b) demande que le *si* soit posé en début de phrase.

(1) Le statut vague de la notion de sujet (et la nécessité de la remplacer par des articulations de concepts opérationnels, tels que ceux de "source", "terme de départ") est frappant dans cet exemple: dans
il vetro si lavora
nous ne saurions dire si *vetro* est sujet ou complément.

NOTE

Dans cette théorie on considère une relation primitive sur les notions qui serait sous-jacente aux énoncés d'une même famille paraphrastique. Sur cette relation s'effectuent des opérations sémantiques qui construisent les différents signifiés des énoncés de la famille.

Toute relation primitive est ordonnée (on a soit: *Paolo picchia Pietro*, soit: *Pietro picchia Paolo*). Mais, à une même relation primitive, on peut faire correspondre des relations prédicatives qui, elles, seront orientées différemment. Ainsi, l'orientation des relations prédicatives qui constituent les signifiés des deux énoncés suivants est différente, mais la relation primitive reste la même: *Paolo batte Pietro; Pietro viene battuto da Paolo*. Sur une relation prédicative, orientée, on fait l'opération du choix du terme de départ. Celle-ci consiste à vider une des places de la relation qui, étant ainsi non-saturée, peut être maintenant considérée comme un prédicat unaire. Le terme qui est choisi comme terme de départ est l'argument de ce prédicat: la relation argument-prédicat est assurée par l'opérateur de repérage $\in$, qui a été défini dans le souci de rassembler les traits des relations exprimées dans les langues par les copules ou leurs substituts et qui paraissent susceptibles de généralisation.

L'ordre de la relation primitive déterminant une "source" et un "but" (respectivement *Paolo* et *Pietro* dans *Paolo batte Pietro*), le passif peut se définir comme l'énoncé dans lequel le terme de départ est le but et l'orientation est but-source.

---

(1) Julien a été blessé
(2), (8): Julien est respecté (**glose:** Julien est respecté par tous)
(3): Ces choses, on ne le fait pas
(4): on travaille
(5): on travaille le verre
(6), (7): Julien est suivi (**glose:** Julien est suivi par qui le suit)
(9): le téléphone est cassé
(10): le téléphone est en train d'être cassé
(11): le verre se travaille
(12): on regarde la reine
(13): la reine se regarde
(14): on peut bien travailler le verre
(15): le verre est travaillé (on le travaille)

(b): **glose:** quelques soient les conditions et les sujets: c'est qui travaille le verre qui travaille le verre et qui travaille le verre, travaille le verre.

ZUR MEHRFACHEN VORFELDBESETZUNG IM DEUTSCHEN

Marc Van de Velde

0. Die Auffassung, im Aussagesatz könne vor dem Finitum nur eine Konstituente stehen, Topikalisierung könne also umgekehrt als Konstituententest verwendet werden, ist weit verbreitet. Sie findet sich z.B. auch noch bei HUBER/KUMMER 1974: 100. Daß sie aber nicht aufrechterhalten werden kann, wurde schon mehrmals gezeigt. Im Folgenden wird auf einen Teilaspekt, die mehrfache Vorfeldbesetzung, näher eingegangen. Aufbauend auf frühere Arbeiten (ENGEL 1970, BENEŠ 1971, FAUCHER 1977) wird versucht, eine Typologie dieses Phänomens auszuarbeiten.
Ein paar Scheinausnahmen werden dabei aber von der Beschreibung ausgeschlossen.
Die nebenordnende Konjunktion bildet, wie H. Brinkmann es ausdrückt, eine Satzbrücke, steht also zwischen den Sätzen, die sie verbindet, und nicht im Vorfeld des nebengeordneten Satzes, verursacht also auch keine mehrfache Vorfeldbesetzung.
Konstituenten können komplex sein, bilden aber, auch wenn der erweiternde Teil relativ frei verschiebbar ist, eine Einheit. Deshalb wird bei Erweiterung durch Appositionen, Attributsätze, nebengeordnete Satzglieder, Adverbien wie _jedoch_ u.ä. keine mehrfache Vorfeldbesetzung angenommen.

1. Bei der eigentlichen mehrfachen Vorfeldbesetzung im Aussagesatz sind 6 große Typen zu unterscheiden.

1.1. Dem eigentlichen Vorfeldelement kann ein anderes Element vorangehen.

Vor dem Vorfeld tauchen relativ frei verschiebbare Elemente auf, die "außerhalb der Subjekt-Prädikat-Beziehung... zu stehen scheinen", wie Anrede, Ausruf und Elemente, die als "Bemerkungen des Sprechers über den Charakter seiner Aussage" (BRINKMANN 1971: 475-476) zu beschreiben sind. Die beiden erstgenannten Kategorien sind, wo sie auch im Satz stehen, deutlich herausgehoben und damit den Parenthesen ähnlich. Bei der letztgenannten Kategorie dagegen braucht das nicht der Fall zu sein. Vgl. die Sätze (1), (2) und (3):

(1) Jedenfalls: diese Sängerin, die mit dem Vornamen Karin hieß, oder Kerstin oder Kirsten oder Karstin oder Karsten, war in Hamar geboren (WH 10)[1]

(2) Das ist jedenfalls das Schema (WH 10)

(3) Jedenfalls kommt sie aus dem bösen Zirkel nicht mehr heraus (WH 12)

Hier stellt sich aber die Frage, ob mit der Voranstellung in (1) nicht auch eine Verselbständigung einhergeht und ob damit nicht wiederum ein Bedeutungswechsel zusammenhängt.

Eine doppelte Schwierigkeit entsteht, wenn das Element herausgehoben zwischen Vorfeldelement und Finitum steht, wie in:

(4) Hochwürden Gusewski, jedenfalls, war immer wieder froh, wenn ich ihm als einer der beiden Ministranten zur Seite stand (GG 46)

Vgl. O. und 1.3.

Eine besondere Stellung nehmen die vorangestellten Nebensätze ein, die keine Inversion bewirken. Bei bestimmten Typen, wie dem je ... desto Satzpaar wird die doppelte Vorfeldbesetzung syntaktisch festgelegt, bei anderen, z.B. beim Konditionalsatz, scheint Schwanken zu bestehen. Ob aber, wie FAUCHER 1977: 28 behauptet, die Drittstellung des Finitums zu einer Interpretation des Satzes als Konzessivsatz führen kann und (5) und (6):

(5) Gelänge ihm dies, nichts wäre gewonnen

(6) Selbst wenn ihm dies gelänge, wäre nichts gewonnen

also synonym sind, scheint uns zweifelhaft.

Als bewußte Abweichung zu betrachten ist die Voranstellung des ohne weiteres zum folgenden Satz gehörenden Satzglieds in:

(7) Trotz Umsturz und Wandel, es verbleiben historische Konstanten (GK 63)

1.2. Als Sonderform der Herausstellung vor das eigentliche Vorfeldelement ist die Linksversetzung (left dislocation) zu nennen:

(8) Aber Celestina, die gibt nichts her (WH 13)

Obwohl diesem Phänomen in der Literatur, vor allem in der transformationell-generativen, schon viel Aufmerksamkeit gewidmet worden ist, bleiben für die Beschreibung doch noch einige Fragen offen, z.B., was als Verweiswort auftreten kann. Normalerweise wird die Möglichkeit beschränkt auf Personal- und Demonstrativpronomina, Pronominaladverbien und wiederaufnehmende Adverbien wie so, da. Hinzugefügt werden müßte auch die Wiederaufnahme durch andere "zusammenfassende" Elemente, wie die Gruppe mit einem Verbalsubstantiv in:

(9) Ob Untersuchungen sinnvoll sind, deren Ziel z.B. das System der Diskussionssprache von Studenten oder Arbeitern ist, die eine moselfränkische Mundart sprechen, eine solche Frage ist nicht von vornherein zu bejahen (Muttersprache 81 (1971): 173)

Auch die Wiederaufnahme durch reine Wiederholung müßte wohl zur Linksversetzung gerechnet werden:

(10) Früher, als ich noch in der Stadt wohnte, und in Deutschland, früher habe ich nachts hin und wieder gern im Telefonbuch gelesen (WH 15)

1.3. Zwischen das eigentliche Vorfeldelement und das Finitum wird ein Element eingeschoben. Es ist noch weiter nach der Form der Parenthese zu unterscheiden.

Eine nichtverbale Parenthese scheint sich vor allem hinter einer Temporalbestimmung zu finden:

(11) Und manchmal, plötzlich, zieht ein jäher Sog von Luft durch die Zimmer, Wind, ein Stoß geballter Luft (WH 5)

Bei einer verbalen dagegen wird das Vorfeld auch oft durch andere Elemente besetzt. Der Einschub kann sowohl Hauptsatz- als Nebensatzfunktion haben.

Der Hauptsatz ist vollständig oder unvollständig. Er kann ein Element enthalten, das auf das Vorfeldelement zurückweist; das braucht aber nicht unbedingt der Fall zu sein:

(12) Nach einer Viertelstunde - Winter schluchzte noch immer - war er wieder auf der Brücke (GG 57)

Vergleicht man:

(13) Diese Sängerin nun, so scheint mir, fügt zu den geringen Eigenschaften des Ortes doch noch eine hinzu, allerdings keine gute (WH 10)

in dem so den ganzen Hauptsatz wiederaufnimmt, mit:

(14) Schließlich - finden Sie - halten sich alle für so unersetzlich, wie er sich hält (HB 229)

so zeigt sich ein deutlicher Parallelismus zu den unter 1.2. besprochenen konditionalen Nebensätzen mit und ohne Verweiswort. (13) wäre also in gewissem Sinne als ein Sonderfall von Linksversetzung zu betrachten, in dem aber nur ein Teil des Wiederaufgenommenen herausgestellt wird.

Besondere Aufmerksamkeit verdient auch der folgende Fall:

(15) Mein Talent, und darin bin ich genial, ist die Bosheit (JL 120)

Er könnte so interpretiert werden, daß der nebengeordnete, durch und eingeleitete Hauptsatz nach vorn gerückt wird, in den ihm vorangehenden Satz hinein.

Der eingeschobene Nebensatz hat oft dasselbe Subjekt wie der Hauptsatz; dieses geht ihm dann auch meistens voran:

(16) Und dieser, nachdem er meine Arbeiten studiert hatte, sicherte mir zu, daß ich in seine Klasse eintreten dürfe (PW 112)

In solchen, oft als Latinismen verpönten, Sätzen ist der Nebensatz einigermaßen mit einem Relativsatz vergleichbar (vgl. FAUCHER 1977: 36). Daß das nicht generell gilt, zeigt:

(17) Die Hand, als könnte ihm der Inhalt wegfliegen, hielt er auf dem Koffer (JL 123)

der aber auch eher abweichend wirkt.

Wenn eine Angabe den Satz eröffnet, ist nicht immer zu entscheiden, ob dieses Element dem Vorfeldnebensatz vorangeht oder ob umgekehrt der Nebensatz eingeschoben ist. Das ist z.B. der Fall in dem von FAUCHER 1977: 47 zitierten Musil-Satz:

(18) Trotzdem, sobald er voll von Eindrücken und Plänen, die vielleicht so reif und neu waren wie nie vorher, zu Hause anlangte, ging jetzt eine entmutigende Veränderung in ihm vor

1.4. Kombination von Elementen wäre anzunehmen in Sätzen wie:

(19) Am Sonnabendmorgen im frühesten Zug war er unausgeschlafen genug, das ganze Unternehmen zu verfluchen (UJ 34)

Auf den ersten Blick könnte das Vorfeld auch so interpretiert werden, daß im frühesten Zug aus dem verkürzten Relativsatz den er im frühesten Zug verbrachte entstanden wäre und also ein Attribut des Vorangehenden bildete. Eine solche Interpretation scheint aber schon nicht mehr möglich im sonst identisch aufgebauten Vorfeld in:

(20) Am Nachmittag in einem Gang zwischen Ladenhäusern ließ B. sich das Flugticket geben und einen geknickten handfleckigen Briefumschlag, der sich dünn anfühlte (UJ 34)

Doch scheinen die beiden Elemente sich durch die gemeinsame Vorfeldstellung näher aufeinander zu beziehen, als wenn sie zusammen im Mittelfeld auftreten. Sie könnten als Glieder eines zusammengezogenen Temporalsatzes (in (19): Als er am Sonnabendmorgen im frühesten Zug saß) aufgefaßt werden. Ob daraus geschlossen werden sollte, daß sie 1 Konstituente bilden, bleibt aber fraglich.

Ähnliches findet sich in:

(21) Nach meiner Meinung am explizitesten hat dies Bolinger schon 1961 unternommen (Lingua 38 (1968): 297)

(22) Nichts mit derartigen Entstehungstheorien hat es natürlich zu tun, wenn... (K. Fleischmann, Verbstellung und Relieftheorie, München 1973: 72)

Hier wird die (objektive oder subjektive) nähere Präzisierung, die sich eigentlich auf den ganzen Satz (oder doch wenigstens auf die Verbindung des anderen Vorfeldelements mit dem Verb) bezieht, in eine besondere Beziehung zu diesem anderen Vorfeldelement gesetzt, ohne daß wiederum von einer eigentlichen syntaktischen Verschiebung von Satzglied zu Satzgliedteil die Rede sein kann.

1.5. Eine besondere Kombinationsform von Vorfeldelementen ist die mit einem Element, das normalerweise rahmenbildend wirkt.

Wenn es sich um ein erweitertes verbales Rahmenelement handelt, wird die Konstruktion allgemein akzeptiert:

(23) Darin eingeschreint war sie in ihrer mageren, märkischen Blüte (GK 111)

Problematisch bleibt dabei aber noch, ob in solchen Sätzen die Regel der Zweitstellung des Finitums im Aussagesatz durchbrochen wird. BAKKERs für Niederländische formulierter, aber ohne weiteres aufs Deutsche zu übertragender Vorschlag, die Kombination einmal als 1 Einheit, zum andern als aus mehreren Gliedern bestehend zu interpretieren (1974: 269), kann auf jeden Fall nicht befriedigen.

Verschieden beurteilt werden dagegen Kombinationen mit einem "inneren Rahmenelement" (cfr. ENGEL 1970: 37-38).
Handelt es sich um ein mit anderen Elementen kombiniertes Adjektiv als Subjekts- oder Objektsprädikativ, dann wird nur von BUNGARTEN 1973: 47 die Grammatikalität bezweifelt:

(24) Albern an ihm fand ich, daß er schon Anfang vierzig einen Antrag auf Namensänderung stellte (GG 89)

Wird im Vorfeld ein dem Verbzusatz nahestehendes Richtungsadverb mit einem anderen Element kombiniert, dann wirkt das schon ungewöhnlicher:

(25) Damit einhergeht häufig im Urteil über sprachliches Verhalten zu kompensierende soziale Verunsicherung, vor allem mittlerer Schichten (Folia Linguistica 5 (1971): 38)

Nicht mehr als normgerecht betrachtet wird meistens die Kombination, in der ein anderes inneres Rahmenelement auftritt, wie in einem Beispiel aus ENGEL 1970: 81:

(26) Die Kinder nach Stuttgart sollst du bringen

Vgl. ERBEN 1972: 270, BUNGARTEN 1973: 31.

1.6. Zum Schluß ist auch noch die Reihung von Elementen im Vorfeld zu beschreiben. Es könnte dabei noch danach unterschieden werden, ob die gereihten Elemente durch Satzzeichen verbunden werden oder nicht.

Keine Satzzeichen werden verwendet in:

(27) Aus unbestimmter Ferne her mit müden Schwingen kam Musik geflogen (H. Hesse, nach BRINKMANN 1971: 478)

Dieser Typ unterscheidet sich von dem vorangehenden (Kombination) darin, daß zwischen den gereihten Elementen nicht der inhaltliche Zusammenhang besteht, der bei den kombinierten zu erkennen ist.

Noch deutlicher tritt der Unterschied zu der Kombination von Elementen hervor, wenn Satzzeichen verwendet werden:

(28) Von dort aus, von jener Minute an, liefen wir durch unbelebte Straßen (GG 122)

Der Unterschied zum Einschub eines Elements nach dem Vorfeldelement (1.3.) liegt darin, daß hier die zwei Vorfeldelemente als gleichwertig aufzufassen sind und das zweite nicht aus dem Satz herausgehoben wird.

Noch mehr als bei den nicht durch Satzzeichen verbundenen Fällen könnte hier übrigens von Nebenordnung ungleichartiger Elemente die Rede sein. Vgl.:

(29) Widerwillig, und nachdem er mehrere Male abgewinkt hatte, stieg Mahlke in die warme Brühe zwischen Strand und erster Sandbank (GG 54)

2. Im Folgenden wird durch einen Vergleich mit der Vorfeldbesetzung in anderen Hauptsatztypen und mit syntaktischen und Wortstellungsphänomenen im Nebensatz versucht, zu einem tieferen Einblick in die oben behandelte Problematik zu kommen.

2.1. Bei den anderen Hauptsatztypen wird der Imperativsatz durch seine Zwischenposition zwischen den Satztypen mit Erst- und mit Zweitstellung des Finitums (vgl. GLINZ 1971: 92) außer acht gelassen und beschränke ich mich also auf die Fragesätze.

Aus einer Untersuchung, was in der Ergänzungsfrage vor dem Fragewort, in der Entscheidungsfrage vor dem Finitum stehen kann, ergibt sich, daß dies nur möglich ist für die dem Aussagesatzvorfeld vorangestellten Elemente, einschließlich der linksversetzten Elemente, und in der Ergänzungsfrage auch für die verbalen Parenthesen vom Typ _(so) scheint mir_. Es müssen aber ein paar Bemerkungen gemacht werden.

Die Selbständigkeit der parenthetischen Adverbien scheint bei Fragesätzen größer zu sein als bei Aussagesätzen: vgl. (1) mit:

(1a) Jedenfalls: wo war diese Sängerin... geboren ?

Ausgeschlossen ist die Voranstellung beim Satzpaar _je_ ... _desto_: das hängt damit zusammen, daß der Nebensatz nur im Vorfeld stehen kann, wenn auch sein Korrelat dem Finitum vorangeht, und das ist im Fragesatz unmöglich.

Bei den anderen Nebensätzen, die keine Inversion verursachen, scheint Wiederaufnahme erforderlich oder doch wenigstens erwünscht zu sein:

(5a) Gelänge ihm dies, wäre dann nichts gewonnen ?

Die übrigen Fälle der doppelten Vorfeldbesetzung tauchen bei den Fragesätzen nicht auf.
Für die Einschübe ist das bei der Entscheidungsfrage leicht zu erklären: dadurch, daß dem Finitum hier nichts vorangeht, geht einfach auch die Möglichkeit des Einschiebens verloren. Bei der Ergänzungsfrage besteht diese Möglichkeit theoretisch wohl; mit Ausnahme des oben behandelten Falls, der auch so unterpretiert werden könnte, daß dem Finitum des Hauptverbs ein Teil des untergeordneten Satzes vorangeht, scheint die Verbindung von Fragewort und Einschub aber unnatürlich:

(11a) Wann, plötzlich, zieht ein jäher Sog von Luft durch die Luft ?

Bei den anderen Fällen (Kombination (auch mit Rahmenelement), Reihung) liegt der Sinn der gemeinsamen Vorfeldbesetzung eben darin, daß eine nicht syntaktisch, sondern kommunikativ zu beschreibende Einheit entsteht. Dieser Sinn geht bei den Fragesätzen verloren, und mit ihm also auch die Möglichkeit, ein Element im Vorfeld (oder vor dem Fragewort) zurückzulassen.

2.2. Bei einem Vergleich mit dem Nebensatz fällt auf, daß bestimmte Elemente, die im Aussagesatz bei mehrfacher Vorfeldbesetzung dem Finitum vorangehen können, überhaupt nicht im Nebensatz auftreten können: es handelt sich um Anrede und parenthetische Adverbien einerseits, durch eine nebenordnende Konjunktion eingeleitete eingeschobene Hauptsätze andrerseits. Bei der ersten Gruppe liegt der Grund in der typischen Leistung dieser Elemente im Kommunikationsakt. Das erklärt auch, warum sie wohl in einem Nebensatz mit Hauptsatzform vorkommen können, wie auch die unten beschriebene Linksversetzung in einem Satz wie:

(30) Ob allerdings der "Mann von der Straße", wie es die Organisatoren erhoffen, auf diese Weise für das Buch gewonnen werden kann, daran bleiben Zweifel, heißt es im Bericht des Börsenblatts für den Deutschen Buchhandel (Kulturbrief 8-9 (1977): 8)

in dem der Hauptsatz aber auch als Endparenthese betrachtet werden könnte.

Daß ein durch eine nebenordnende Konjunktion eingeleiteter Zwischensatz im Nebensatz nicht auftreten kann, ist aus seinem möglichen Ursprung zu erklären (vgl. 1.3.).
Problematisch ist, ob die Linksversetzung sich auf Hauptsätze beschränkt. Die von HIRSCHBÜHLER 1974: 15 aufgeführten französischen Beispiele mit dieser Konstruktion im Nebensatz, wie:

(31) je sais bien que des pommes, tu ne lui en as donné que deux

können auf jeden Fall nicht direkt ins Deutsche umgesetzt werden. Ein Grund dafür kann vielleicht sein, daß das Deutsche, anders als das Französische, eine eigene Nebensatzform hat.
Im einzigen in Betracht kommenden Fall, den ich fand, weist zwar das wiederaufnehmende <u>es</u> auf die hier behandelte Konstruktion hin, die Stellung des Objektsatzes nach dem Subjekt erlaubt es aber nicht, von Linksversetzung zu reden:

(32) daß er, was er auch getan haben mag, es aus Überzeugung getan hat (Die Zeit Nr. 30/1967: 28, nach Pütz H.: Über die Syntax der Pronominalform <u>es</u> im modernen Deutsch. Tübingen (1975): 158)

Beschränkt ist auf jeden Fall auch die Möglichkeit, im Nebensatz vorzukommen, für Zwischensätze vom Typ <u>(so) scheint mir</u>. Sie scheinen beschränkt auf bestimmte Typen von Nebensätzen der indirekten Rede, wie:

(13a) Er will nicht kommen, da, so scheint ihm, diese Sängerin zu den Eigenschaften des Ortes doch keine mehr hinzufügt

Die anderen Verbindungen von Elementen, die zusammen das Vorfeld des Aussagesatzes besetzen können, scheinen ohne weiteres auch in Nebensätzen vorkommen zu können. Ihre Stellung hat dann aber nichts mehr mit den Regeln für die Vorfeldbesetzung zu tun, sondern wird durch die Regeln bestimmt, die die Distribution im Mittelfeld beschreiben; d.h., daß im Nebensatz solche (Verbindungen von) Elemente(n) der Subjunktion unmittelbar folgen können, die im Hauptsatz unmittelbar hinter dem Finitum stehen können. Unmöglich ist das z.B. für das <u>je</u> ... <u>desto</u> Satzpaar oder für die Verbindung mit einem Rahmenelement. Die Regel, daß Subjunktionen möglichst selten nebeneinander vorkommen sollten, wird am leichtesten bei einem untergeordneten Konditionalsatz durchbrochen, was der Forderung nach einer logischen Reihenfolge (Voraussetzung-Vorausgesetztes) entspricht Daraus wäre dann vielleicht die Voranstellung, auch vor die Subjunktion des übergeordneten Nebensatzes, also der einzige Fall eines Nebensatzes mit einem Vorfeld, zu erklären:

(33) weil er glaubt, wenn der Körper bei Seite geschafft sei, daß sodann nichts seine Tat verraten könne (Lessing, nach Behaghel, O.: Deutsche Syntax IV, Heidelberg (1932): 283)

Diese Wortstellungserscheinung wurde, soweit ich weiß, in einem transformationellen Rahmen noch überhaupt nicht behandelt. Ist sie als ein weiterer Fall von Hebung (raising) zu beschreiben ?

3. Aus unserem Vergleich der mehrfachen Vorfeldbesetzung im Aussagesatz mit den Stellungsmöglichkeiten in anderen Hauptsatztypen und im Nebensatz ergibt sich eine Zweiteilung. Zur ersten Gruppe gehören die Anrede und die parenthetischen Adverbien: sie kommen auch vor einem Fragesatz, nicht aber im Nebensatz vor. Die erste Eigenschaft teilen mit ihnen die linksversetzten Elemente, die zweite (wahrscheinlich) nicht. Die Einschübe bilden eine uneinheitliche Gruppe, die noch weiter eingeteilt zu werden braucht. Allen den genannten Elementen ist weiter noch gemeinsam, daß sie nicht allein im Vorfeld auftreten. Darin unterscheiden sie sich von den anderen Verbindungen, die nicht im Fragesatz, wohl aber im Nebensatz vorkommen können. Hieraus ergibt sich, daß die Beschreibung der mehrfachen Vorfeldbesetzung nicht ein alleinstehendes Problem ist, sondern mit anderen syntaktischen und topologischen Phänomenen zusammenhängt und also tieferliegenden Gesetzmäßigkeiten unterworfen ist, als man zunächst denken könnte.

ANMERKUNG

1 Auf mehrmals zitierte Quellen wird durch eine Abkürzung hingewiesen. WH: W. Hildesheimer, Tynset: Fischer Taschenbuch 844: 1970. GG: G. Grass, Katz und Maus: rororo 572: 1963. HB: H. Böll, Ansichten eines Clowns: DTV 400: 1970. GK: G. Kunert, Im Namen der Hüte: DTV-Sonderreihe 88: 1970. JL: J.

Lind, Landschaft in Beton: Neuwied, Luchterhand: 1963. PW: P. Weiß, Abschied von den Eltern: edition suhrkamp 85: 1969. UJ: U. Johnson, Zwei Ansichten: rororo 1068: 1968. Der Abkürzung folgt die Seitenzahl.

LITERATUR

BAKKER, Dirk M. (1974): "Zinnen als reeksen: een vergelijkende standpuntbepaling". De nieuwe taalgids 67: 265-278.

BENEŠ, Eduard (1971): "Die Besetzung der ersten Position im deutschen Aussagesatz". Fragen der strukturellen Syntax und der kontrastiven Grammatik. Düsseldorf: Schwann. (= Sprache der Gegenwart 17)

BRINKMANN, Hennig (1971): Die deutsche Sprache. Gestalt und Leistung. Düsseldorf: Schwann, 2.A.

BUNGARTEN, Theo (1973): Umstellprobe und Minimalrepräsentierung als Methoden der Satzanalyse. BUNGARTEN, Theo (ed.): Sprache und Sprachanalyse des Deutschen. Bern etc. Lang: 9-72.

ENGEL, Ulrich (1970): Regeln zur Wortstellung. ENGEL, Ulrich (ed.) Forschungsberichte des Instituts für deutsche Sprache 5. Mannheim: 9-148.

ERBEN, Johannes (1972): Deutsche Grammatik. Ein Abriß. München: Hueber, 11.A.

FAUCHER, Eugène (1976): "Exceptions à la règle V2 ?". Cahiers d'Allemand 11: 27-47.

GLINZ, Hans (1971): Deutsche Grammatik I: Satz-Verb-Modus-Tempus. Frankfurt: Athenäum.

HIRSCHBÜHLER, Paul (1974): "La dislocation à gauche comme construction basique en français". ROHRER, Christian / RUWET, Nicolas: Actes du Colloque Franco-Allemand de Grammaire Transformationelle I: Etudes de Syntaxe. Tübingen: Niemeyer: 9-14. (= Linguistische Arbeiten 13)

HUBER, Walter / KUMMER Werner (1974): Transformationelle Syntax des Deutschen I. München: Fink.

ON THE SIMPLICITY CRITERION IN LINGUISTICS

Wolfgang Grassl

As can easily be realized from the abundant literature on this issue, the structure and the status of the simplicity criterion is one of the most persistent and controversial topics in the methodology of linguistics of the past two decades. Needless to say, simplicity considerations have played a major role particularly in work accepting the transformational-generative model as a framework for linguistic inquiry, without having been confined, however, to that particular research program. Appeals to simplicity can be found already in the earliest of Chomsky's published works, though the theoretical justification of the simplicity criterion has since changed considerably. While in his earlier works he employed a purely formal notion of simplicity, having been inspired by Goodman's and Quine's nominalistic views, Chomsky's 'standard theory' was intended to empirically motivate the concept by attributing a certain psychological reality to the simplest among rival descriptions. In the third, more defensive phase, in which he developed his 'extended standard theory', Chomsky also revised his views on simplicity considerably, no longer conceiving of a simpler solution as preferable in every case.[1]

One of the basic tenets common to all scientific methodologies in which simplicity considerations have any relevance at all is that simplicity is at least one factor to be considered when choosing between competing hypotheses or theories. This is, however, about as much unanimity as can be found among 'orthodox' philosophers of science on this issue.[2] The reasons given as to why simplicity should play any role at all in the assaying of scientific theories diverge considerably according to the respective metascientific points of view. One argument that has been proposed for justifying this appeal to simplicity as an evaluation standard will be scrutinized here. We shall, however, not be concerned with the p s y c h o l o g i c a l motives a scientist might have for resorting to the simplest of a given set of hypotheses (cf.QUINE 1966/rev.ed.1976:256f.), but shall try to inquire on what *objective* grounds such a preference might be based. Concerning this question, no unanimous answer can be obtained from contemporary philosophy of science and epistemology. Although simplicity is often considered to be a compelling factor in theory choice, its nature has yet remained opaque.

It appears that, on the most general level, we can distinguish two basic standpoints that have been taken in the philosophy of science regarding the simplicity criterion. Accounts that link the simplicity of a hypothesis or theory with its probability can be distinguished from nonprobabilistic proposals, such as esthetic or pragmatic considerations. Probabilistic accounts have again come in generally two variants. We shall call these the *high probability view* and the *low probability view*. Among these only the latter has been claimed for linguistics and it is only to this conception that we shall direct our attention.

Philosophers advocating a low probability view of simplicity maintain that the stronger and thus less probable and more falsifiable of two competing hypotheses is the simpler one. Popper connects the notion of simplicity with the notion of *empirical content*, i.e.with the amount of information a hypothesis contains about the world. Thus Popper claims that a hypothesis is the simpler the more empirical content it has, and this again he identifies with the degree of falsifiability of the hypothesis (cf.POPPER 1934/5th ed.1973:102f.). Popper's justification of the simplicity criterion therefore basically consists in asserting that simpler hypotheses are to be preferred to less simple ones since they are more informative, easier falsifiable and have a higher empirical content.

Since the Popperian approach has been repeatedly claimed as methodological framework for linguistics, it will deserve some further scrutiny.[3] Instead of discussing objections that have been proposed within general philosophy of science, we shall, however, rather look at which problems are involved when applying the low probability view developed for theory-choice in the natural sciences to the selection among competing linguistic hypotheses. Thus we shall have to specify criteria by which two hypotheses may be considered as competing in the first place.

Hardly any metatheoretical notion is used by linguists more equivocally than the term 'hypothesis', which is sometimes employed in such a wide sense as to cover almost any proposal put forth by a linguist. Rather than engaging in the task of formally defining the concept of hypothesis for linguistics, it seems in this case advisable to give this notion quite a broad range of application, so as not to have to demarcate sharply between hypotheses and laws. We shall conceive of this difference as rather of an epistemic than of a purely logical nature, disregarding for the present purpose the very controversial question of whether linguistic rules like extraposition, Equi-NP-deletion, wh-movement or any phonological rule may be regarded as laws in the sense of the natural sciences. We shall rather speak of hypotheses on different levels, lower-level hypotheses having the form of singular or existential sentences, higher-level hypotheses being of a more universal and nomological character (cf. BUNGE 1967a:222ff.).

It is a common assumption in philosophy of science that two hypotheses H_1 and H_2 are *competing* or *alternatives* if they both suggest nonequivalent answers to a common problem. Regarding H_1 and H_2 as competing presupposes that they are *commensurable*, i.e. that there exists at least one procedure by which H_1 and H_2 can be compared according to a common standard (cf. BUNGE 1967a:258). It furthermore presupposes that H_1 and H_2 are *incompatible*, that is to say, that they cannot be asserted jointly.

What researchers in the natural sciences usually assume when speaking of competing hypotheses is that they are dealing with a set of accurately testable hypotheses. Only hypotheses that have been found by usual test procedures to be empirically equivalent may be evaluated according to their respective simplicity. Two hypotheses are *empirically equivalent* if in addition to being rival they yield much the same testable consequences or predictions, that is, they have the same empirical content. Empirical equivalence does not, however, presuppose mutual compatibility; various cartographic representations of the globe can thus be empirically equivalent but incompatible. If empirically equivalent hypotheses are also mutually compatible, they would be just different formulations of the same hypothesis, i.e. c o n c e p t u a l l y e q u i v a l e n t (cf.BUNGE 1967b:347).

It is our contention that what linguists regard as competing hypotheses are in a large number of cases conceptually equivalent hypotheses. All the well-known procedures where the relative merit of a proposed phonological or syntactical rule is ascertained by counting the number of symbols it contains and then selecting the more parsimonious and thus simpler one can only be applied to conceptually equivalent hypotheses. Notational conventions such as Halle's feature counting (cf.HALLE 1962:54ff.), the use of numerical subscripts, curly brackets or variable coefficients in generative phonology, are based on what in the philosophy of science has been called n o t a t i o n a l s i m p l i c i t y. Moreover, proposals like Chomsky's and Halle's suggestion to measure the complexity of a rule by the number of symbols it contains, i.e.by its length (cf.CHOMSKY 1965:42), may be said to rely on the notion of o b j e c t i v e-n o t a t i o n a l s i m p l i c i t y , since it does not involve subjective-psychological claims like the familiarity of notation, its convenience etc. (cf.RUDNER 1966:43). Notationally different hypotheses in linguistics would, however, hardly be regarded as competing in any of the natural sciences, since it cannot reasonably be held that two differently formulated rules describing, for instance, palatalization in Russian, the Great Vowel Shift in English, or the Italian diphtongization process that leads to 'dittonghi mobili' in open syllables, are mutually incompatible. The only other sciences where the conception of objective-notational simplicity is employed quite frequently are what has come to be called the 'formal sciences', i.e.mathematics and logic. The basic rationale behind coalescing two phonological rules into one and thus having to use fewer symbols is the same as behind simplifying a logical formula by transforming it into its disjunctive normal form or behind the rules of cancellation in arithmetic. While in arithmetic and logic, however, simplification of notation is a device for making equations easier computable or, in the case of logic, for comparing formulae in order to determine whether they are tautologically equivalent, generative linguists like Chomsky claim that notational conventions have empirical import. It is argued that "a person learning a language will attempt to formulate generalizations that can easily be expressed (that is, with few symbols) in terms of the notations available ... , and that he will select grammars containing these generalizations over other grammars that are also compatible with the given data, but that contain different sorts of generalizations ..."(CHOMSKY 1965:45).

Given the present state of psycholinguistic research it seems, however, justified to say that no such claim can yet be substantiated by empirical test procedures. Chomsky introduced his assumption that children learning a language will select the grammar allowing the greatest number of generalizations and will internalize the simplest set of rules just in order to account for the amazing ability of a child to learn the complex mechanisms and structures of a natural language in a relatively short period. This psychological claim has the status of an auxiliary hypothesis intended to back up the assumption of the empirical status of notational conventions. Whereas it is usually required of auxiliary hypotheses that they be independently empirically justified, i.e.independently testable so as to ascertain their truth-content, no test procedures have been devised yet to substantiate Chomsky's psychological assumption.

Pertinent psycholinguistic studies would, quite on the contrary, rather supply arguments against this assumption. It is sufficient to refer in this respect to the failure of the attempts at validating the 'derivational theory of complexity' which postulated a correlation between linguistic economy, mirrored by such formal features as the length of rules, and performative parameters, such as perceptual difficulties, ease of recall etc.[4] It thus seems quite reasonable to agree with the conclusion drawn by Bever, that "the failure of linguistic derivational complexity to predict psychological complexity demonstrates that the relation between current grammars and perception is remote if it can be said to be extant at all" (BEVER 1968:57). Furthermore, Chomsky's auxiliary hypothesis has recently come under attack in the field of child language acquisition. On the basis of empirical investigations it is argued that "there is no form of an evaluation measure which can be said to play a role in CLA" (CHURMA 1975:98). Not wanting to judge the validity of this claim, it may suffice to point out here that work in psycholinguistics and child language acquisition has tended rather to contradict Chomsky's psychological assumption than to corroborate it.

What this amounts to is n o t the requirement that linguists should not prefer among conceptually equivalent rules the one that makes use of fewer symbols, but rather that claims regarding the psychological reality of notational conventions should be given up as long as they cannot be motivated by substantive evidence from psycholinguistic research. Simplicity considerations should only be employed for an easier comparison of rules, a simpler application and the like. As has recently been pointed out by a generative phonologist, "most of the arguments in phonology today reduce to arguments over esthetic judgments" (SCHANE 1976:183). From this follows that no probabilistic account of simplicity can justify the adoption of the current abbreviatory conventions in generative grammar, but that these conventions should rather be interpreted as based on an esthetic-pragmatic conception of simplicity.[5]

Let us now look what role simplicity considerations may have in deciding between two higher-level hypotheses or between different theories. Consider as exemples any of the well-known proposals put forward by adherents of the extended standard theory and by generative semanticists concerning the analysis of imperative constructions, the treatment of sentence adverbs etc. How is it possible to assess the relative merits of any of these competing proposals and to ascertain the better one? In order to uphold the analogy with the natural sciences that has been claimed time and again by generative linguists (cf.CHOMSKY / KATZ 1974:356f.), we would have to require that the hypotheses under consideration possess properties which render them empirically testable, at least in principle. Testability means, for philosophers of science as Popper and Hempel, that a hypothesis is susceptible to disconfirmation or refutation. A hypothesis is usually tested by deducing predictions or consequences from it which are subsequently confronted with observational evidence.[6]

This method of testing hypotheses has repeatedly been claimed to be applicable also in generative linguistics. One of its most fervent advocates writes:

> If we are comparing alternative grammars, we can determine via deductive reasoning the empirical claims of each grammar so that we may determine the areas of overlap where the grammars make different claims about the same range of data. (DOUGHERTY 1974:128)

More generally, Dougherty sketches the methodology of linguistic inquiry in the following way:

> The method of theory-construction — the process which starts with observation, goes on to invent hypotheses, formulates the hypotheses as algebraic models and deduces their consequences, and terminates in inductive verification or refutation — is the method of research used in transformational generative grammar. (DOUGHERTY 1976:62)

This program certainly conforms to the 'received' view of scientific inquiry developed in greatest depth for the natural sciences. The crucial difference between linguistics and, on the other hand, the natural sciences does not, however, manifest itself on the level of theory-construction, but already on an earlier and conceptually prior stage: in deciding what ought to count as 'data' and what the notion of 'empiricalness' is to mean in linguistics. It has often been argued by generative linguists that the intuitions of an informant — who may even be identical with the linguist — are to be considered as the observational data in linguistic theory-construction. Consequently, testing a certain hypothesis is usually understood as confronting it with a native speaker's intuitions.

Numerous problems connected with this practice have been pointed out both by linguists and by philosophers of science.[7] Many of the objections boil down to the argument that no e m - p i r i c a l data may be obtained by Chomsky's introspective procedures, thus vitiating comparisons between linguistics and the natural sciences. Empiricalness in this sense may be defined in the Popperian fashion as 'falsifiable on the basis of observational evidence', observable phenomena being conceived of as occurring in space and time. Consequently, if one were to maintain Popper's philosophy of science as a methodology for linguistics, potential falsifiers of linguistic hypotheses would have to be spatiotemporal events or states. Limiting its domain to an investigation of linguistic competence, transformational-generative linguistics clearly cannot be said to be based on empirical data in this sense, and consequently it makes no sense to speak of the empirical content of a hypothesis or theory in generative linguistics, thus failing to comply with the basic assumption of the low probability account of simplicity.

There is obviously no external evidence on which a decision between two higher-level hypotheses might be based. The hypotheses are not empirically testable in the standard sense of the term. Neither could native speaker intuitions help to decide between, e.g., two proposed deep structure representations, nor will simplicity considerations apply to such hypotheses, since it is by no means clear what should be taken as criteria for the respective simplicity of the hypotheses. Requirements like 'do not postulate unnecessary deep structure elements' would be based on a conception of ontological simplicity that would in itself not be susceptible to any independent empirical motivation. All these considerations certainly lend support to the point made by Quine, that there are no natural and nontrivial criteria for deciding between rival grammars (cf.QUINE 1972:443f.). It thus seems legitimate to say that a choice between two hypotheses that are derived from different general theories or even different research programs cannot be made only on the basis of simplicity considerations. This means that, for instance, no single hypothesis of interpretive semantics or of the recently developed trace theory of movement rules can be compared with one stemming from a generative semantic theory by assessing their respective simplicity, without having other evaluation measures to judge between the general theories of interpretive semantics or trace theory, and generative semantics. This conclusion ties in with a widely accepted opinion in philosophy of science:

> Simplicity seems to be largely an intratheoretic matter. It is not just any two hypotheses, but rather two hypotheses within a common framework that stand a chance of being compared in terms of their respective simplicity. (LAMBERT / BRITTAN 1970:72)

Neither does it make sense to compare rules of transformational generative phonology with rules of natural generative phonology by taking simplicity as an evaluation measure, nor to compare an extended standard theory account of some grammatical phenomenon with a corresponding Montague-style analysis by resorting to simplicity considerations, since both hypotheses depend on specific general frameworks, the validity of which has to be ascertained by other criteria than that of simplicity.

Let us briefly sum up our conclusions. The methodological difficulties for a Chomsky-type linguistics are, in the first place, due to its limitation to the study of linguistic competence which involves relying solely on native speaker intuitions in hypothesis- or theory-choice. This methodological orientation makes inapplicable to the field of generative linguistics the simplicity criterion developed by Popper for choosing between competing empirically equivalent hypotheses or theories in the natural sciences. As regards lower-level linguistic hypotheses, notational conventions can be shown to rely on a purely formal, notational conception of simplicity, any claim of their having a psychological import being beyond the range of empirically falsifiable assertions. Less complex formulations should be preferred to more complex ones for purely esthetic or pragmatic reasons.

Regarding higher-level hypotheses or theories, Popper's account was found not to be applicable, either. Since simplicity appears to be to a large extent an intratheoretic measure — as was also realized by Chomsky (cf.CHOMSKY 1965:37ff.) — the application of the simplicity criterion to the case of competing hypotheses derived from different linguistic theories is meaningless. While in the natural sciences two rival hypotheses or theories are subject to empirical testing, and only those hypotheses or theories that have been found to be empirically equivalent will subsequently be evaluated according to their respective simplicity, the mentalistic orientation of generative linguistics with its introspective testing methods makes a far more extensive use of appeals to simplicity without, however, supplying an empirical justification for them.

A way out of this methodological predicament is only conceivable by giving up the limitation to the study of competence and the reliance on intuitions on which generative linguistics is based. Not intending to require that intuitions about grammaticality be completely abandoned as a source of linguistic evidence, as has recently been proposed by some linguists (cf.SAMPSON 1975: ch.4), we want, however, to maintain that intuitions belong rather to the pretheoretical than to the theoretical stage of linguistic inquiry.[8] They may well serve as heuristic, research-guiding tools in the 'context of discovery', but we should be hesitant to rely on them in justifying linguistic hypotheses. This implies a shift of the basic emphasis of linguistic investigation to those aspects of its domain which are capable of having empirical claims made about them. Performance data and behavioral data will gradually have to be given more weight and attention, since the arguments based on internal criteria like simplicity have certainly reached an impasse.[9] If the domain of data used by generative linguists in grammatical analyses were extended to cover also evidence from psycho- and sociolinguistics, child language acquisition, speech errors, neurophysiology, experimental phonetics and the like, it is to be expected that linguistic hypotheses will be more easily susceptible to empirical testing and falsification, thereby attaining empirical content. This might reduce the extensive use of the simplicity criterion in current linguistic practice and assimilate simplicity considerations in linguistics to those in other scientific disciplines.

NOTES

1 For a sketch of the development of Chomsky's views on simplicity see WEYDT 1975b:65-76.

2 We shall not discuss the positions of the influential group of radical meaning variance theorists (Feyerabend, Kuhn, Hanson and others) who deny the possibility of finding rational and nontrivial criteria for comparing different theories. The assumption of the logical incommensurability of rival theories seems to clash with the fact that scientists do in fact make logical and experimental comparisons of such theories adopting certain evaluation criteria.

3 Cf.as a representative example KANNGIESSER 1972:ch.1.3.

4 Cf.the exposition in DERWING 1973:282ff.

5 For a similar conclusion see WEYDT 1975a:314.

6 For such an account see HEMPEL 1966:88f.

7 Cf.BOTHA 1971; RINGEN 1975; LABOV 1975:14-41.

8 Cf.BUNGE 1962:113 who holds that the 'intuitive' stage of a discipline is its presystematic stage.

9 For a similar view see LEHRER 1972.

BIBLIOGRAPHY

BEVER, Thomas G. (1968): "A survey of some recent work in psycholinguistics". PLATH, W.J. (ed.): Specification and Utilization of a Transformational Grammar. Yorktown Heights: IBM Research Center: 1-66.

BOTHA, Rudolf P. (1971): Methodological Aspects of Transformational Generative Phonology. The Hague:Mouton.

BUNGE, Mario (1962): Intuition and Science. Englewood Cliffs: Prentice-Hall.

--- (1967a): Scientific Research I.The Search for System. Berlin etc.:Springer.

--- (1967b): Scientific Research II.The Search for Truth. Berlin etc.:Springer.

CHOMSKY, Noam (1965): Aspects of the Theory of Syntax. Cambridge: MIT Press.

CHOMSKY, Noam / KATZ, Jerrold J. (1974): "What the linguist is talking about". Journal of Philosophy 71: 347-367.

CHURMA, Donald G. (1975): "Child language acquisition and the justification of linguistic hypotheses". GROSSMANN, Robin / SAN, L.James / VANCE, Timothy J. (eds.): Papers from the Eleventh Regional Meeting. Chicago:Chicago Linguistic Society: 92-101.

DERWING, Bruce L. (1973): Transformational Grammar as a Theory of Language Acquisition. Cambridge:Cambridge University Press 1973.

DOUGHERTY, Ray C. (1974): "What explanation is and isn't". COHEN, David (ed.): Explaining Linguistic Phenomena. New York etc.:Wiley: 125-151.

--- (1976):"The Logic of Linguistic Research"(unpublished).

HALLE, Morris (1962): "Phonology in a generative grammar". Word 18: 54-72.

HEMPEL, Carl G. (1966): Philosophy of Natural Science. Englewood Cliffs:Prentice-Hall.

KANNGIESSER, Siegfried (1972): Aspekte der synchronen und diachronen Linguistik. Tübingen:Niemeyer.

LABOV, William (1975): What Is a Linguistic Fact? Lisse:Peter de Ridder Press.

LAMBERT, Karel / BRITTAN, Gordon (1970): An Introduction to the Philosophy of Science. Englewood Cliffs:Prentice-Hall.

LEHRER, Adrienne (1972): "Evaluating grammars:what counts as data?". Lingua 29: 201-208.

POPPER, Karl R. (1934): Logik der Forschung. Wien:Springer. 5th ed.1973. Tübingen:Mohr.

QUINE, Willard Van Orman (1972): "Methodological reflections on current linguistic theory". HARMAN, Gilbert / DAVIDSON, Donald (eds.): Semantics of Natural Language. Dordrecht: Reidel: 442-454.

--- (1966): "On simple theories of a complex world". QUINE, Willard Van Orman: The Ways of Paradox and Other Essays. Cambridge etc.:Harvard Univ.Press. Rev.ed.1976: 255-258.

RINGEN, Jon D. (1975): "Linguistic facts:a study of the empirical scientific status of transformational generative grammar". COHEN, David / WIRTH, Jessica R. (eds.): Testing Linguistic Hypotheses. New York etc.:Wiley: 1-41.

RUDNER, Richard S. (1966): Philosophy of Social Science. Englewood Cliffs:Prentice-Hall.

SAMPSON, Geoffrey (1975): The Form of Language. London:Weidenfeld and Nicolson.

SCHANE, Sanford A. (1976): "The best argument is in the mind of the beholder". WIRTH, Jessica R. (ed.): Assessing Linguistic Arguments. New York etc.:Wiley: 167-185.

WEYDT, Harald (1975a): "Das Simplizitätsprinzip in der Linguistik". Folia Linguistica 7: 307-316.

--- (1975b): "Das Problem der Sprachbeschreibung durch Simulation". SCHLIEBEN-LANGE, Brigitte (ed.): Sprachtheorie. Hamburg:Hoffmann und Campe: 53-80.

3. SEMANTIK

DIREKTIONALE PRÄPOSITIONEN UND KONTEXTTHEORIE

Arnim von Stechow

1. Das Problem +)

Ich interessiere mich auf den folgenden Seiten für einige räumliche Adverbiale wie "durch den Hotzenwald", "um mich im Kreis herum" und "im Zug". Meine Überlegungen sind inspiriert durch das faszinierende Papier "Prepositions and points of view" von Max Cresswell. Dort erhält ein Satz wie

(1) Randi geht durch den Hotzenwald

ungefähr die folgende Wahrheitsbedingung:

1.1 (1) ist wahr in einer Welt w während eines Zeitintervalls z genau dann, wenn Randi in w während z geht und ihr Weg in w während z durch den Ort des Hotzenwaldes in w zu z führt. Bevor ich diese Wahrheitsbedingung kommentiere, erläutere ich zunächst meine Redeweisen im Zusammenhang mit Ortsbestimmungen. Sie sind ziemlich ungewohnt, weil diese Dinge meist vernachlässigt werden. Wegen der Wichtigkeit dieser Begriffe für meine Überlegungen erlaube ich mir aber eine gewisse Pedanterie.

Wie üblich setze ich eine Menge W von möglichen Welten voraus und eine Menge Z von Zeitintervallen.[1] Ferner gibt es eine Menge A von möglichen Dingen (Individuen). Zu jeder Welt $w \in W$ gehört der absolute Raum R, den man sich durch $\mathbb{R}^3$ repräsentiert denken kann, wobei $\mathbb{R}$ die Menge der reellen Zahlen ist. Jeder Raumpunkt von R läßt sich dann identifizieren mit einem Tripel $\langle a,b,c\rangle$ von reellen Zahlen.

1.2 Definitionen (a) Ein Ort ist ein Paar $\langle M,n\rangle$ wobei $M \subseteq R$ und $n \in R$. M heißt Gebiet des Ortes, n sein Nullpunkt. (b) $\langle M_1,n_1\rangle$ ist ein Teilort von $\langle M_2,n_2\rangle$ gdw. $M_1 \subseteq M_2$. (c) Die Bahn des Dinges a in der Welt w während der Zeit z ist diejenige Funktion $B_{a,w,z}$, so daß für beliebige Momente $m \in z$ gilt: das Gebiet von $B_{a,w,z}(m)$ ist der Raum, den a zu m ausfüllt; der Nullpunkt von $B_{a,w,z}$ ist der Raumpunkt, den der "Mittelpunkt" von a in w zu m einnimmt. (d) Der Weg des Dinges a in der Welt w während der Zeit z ist diejenige Funktion $P_{a,w,z}$ so daß für jedes $m \in z$ gilt:

$P_{a,w,z}(m)$ = der Nullpunkt des Ortes $B_{a,w,z}(m)$. (e) Eine *Ortsbestimmung* o für (das Zeitintervall z) ist eine Funktion, die jedem $m \in z$ einen Ort zuordnet. (f) Sei o eine Ortsbestimmung für das Intervall z. Wir sagen: das Individuum *a befindet sich während z am Ort o* gdw. für jedes $m \in z$ $B_{a,w,z}(m)$ ein Teilort von o(m) ist.

Für das folgende gehe ich davon aus, daß es einen *absoluten Ort* , <R,Null>, gibt. R heißt *Weltenraum* und Null absoluter Nullpunkt. Null ist repräsentiert durch <0,0,0>. Sprachlich kann eine Ortsbestimmung durch lokale Adverbiale wie *im Speisewagen* ausgedrückt werden. Für ein Intervall z ist damit der jeweilige Ort gemeint, der durch das Innere des Speisewagens für jeden Moment von z besetzt wird.

Nach diesen Begriffsbestimmungen kommentiere ich nun die Wahrheitsbedingung 1.1. Zunächst eine Bemerkung dazu, was mir daran gut gefällt. Die direktionale Präposition "durch" drückt hier bei genauerem Hinsehen eine Relation zwischen einem Weg und einem Ort aus. Ich glaube, das ist tatsächlich das hervorstechenste Merkmal ihrer Bedeutung. Völlig Analoges gilt auch für andere direktionale Präpositionen. So kann man sich etwa den Sinn der Präpositionen "durch", "im Kreis herum um" oder "über hinweg" sofort durch die folgenden Bildchen veranschaulichen:

 durch  im Kreis herum um über hinweg 

Die Idee des Weges und seiner Form steht bei dieser Analyse ganz in der Präposition und nicht im Verb "gehen". Man kann ja durchaus auf der Stelle gehen. Es ist auch kein Problem, die Bedeutungen für die einzelnen Wörter von (1) so anzusetzen, daß gerade die Wahrheitsbedingung 1.1 herauskommt. Das alles sind schöne Ideen, die ich im wesentlichen beibehalten möchte.

Die Analyse 1.1 ist aber in dieser Form noch nicht korrekt. Nehmen wir nämlich einmal an, Randi gehe während der Zeit z im Speisewagen des Intercityzuges "Räuber Hotzenplotz" um mich im Kreis herum, wobei dieser Zug während z den Hotzenwald durchquert. Nach 1.1 müßte dann (1) während z wahr sein, denn der Weg von Randi zur Zeit z führt durch den Hotzenwald, weil der "Räuber Hotzenplotz" während z den Hotzenwald durchquert.

Diese Konsequenz ist aber absurd. Für die geschilderten Umstände trifft es zu, daß Randi im Kreis herum geht, daß sie durch den Hotzenwald fährt und, in einem sehr physikalischen Sinn von "bewegen", auch: daß sie sich durch den Hotzenwald bewegt. Es trifft aber nicht zu, daß sie durch den Hotzenwald geht.

Ich habe in meinem Papier von Pavia mehrere Möglichkeiten diskutiert diese Schwierigkeit zu überwinden. Inzwischen habe ich sie allesamt verworfen und bin stattdessen zu der Überzeugung gelangt, daß der Fehler der Wahrheitsbedingung 1.1 in der *absoluten* Bestimmung der Bewegungen, d.h. der zurückgelegten Wege der Aktanten liegt. Dies wird besonders deutlich, wenn man etwas kompliziertere Beispielssätze betrachtet wie z.B.:

(2) Während Angelika durch das Zimmer schritt, lief Heathcliff laut kläffend dreimal im Kreis um sie herum.

(3) Im Speisewagen lief Heathcliff schwanzwedelnd dreimal im im Kreis um Angelika herum, während sie ihr Ei aß.

Für die Beurteilung, ob die durch (2) und (3) ausgedrückten Sachverhalte u.U. der Fall sind, kommt es offenbar nur auf den Ort an, *an dem* Angelika und Heathcliff gerade sind. Es spielt keinerlei Rolle, in welcher Weise dieser Ort sich seinerseits wieder bewegt. Wichtig ist, daß Heathcliffs Weg einen Kreis um Angelika beschreibt. Absolut bestimmt ist Heathcliffs Weg aber sicher kein Kreis um Angelika. Im Falle von (2) ist er bereits relativ zum Ort des Zimmers ein Zykloid, also so etwas:

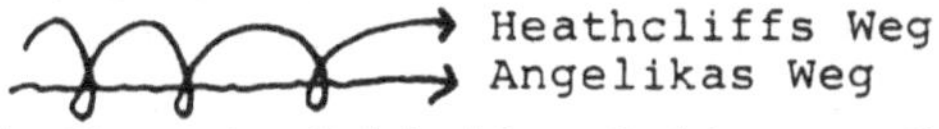

Im zweiten Fall ist Heathcliffs Weg ein Zykloid relativ zum Ort der Bahngeleise,falls der Speisewagen gerade rollt. Angelikas Weg könnte in beiden Fällen eine mehr oder weniger gerade Strecke sein, falls die Erde sich nicht bewegt. Es dürfte ziemlich schwierig(u. vor allem abwegig) sein, die Bedeutung der Präposition "im Kreis herum um" zu beschreiben als ein Verhältnis *solcher* Wege zueinander.

Die Idee, die ich im folgenden entwickele, ist sehr einfach. Cresswells Analyse wird dahin abgeändert, daß die Wege der betroffenen Individuen auf eine Ortsbestimmung hin relativiert werden. Diese Ortsbestimmung kann manchmal teilweise sprachlich explizit gemacht werden. Das leistet beispielsweise im Satz (3) das lokale Adverbial "im Speisewagen". Meistens muß aber aus dem Verwendungskontext eines Satzes heraus klar sein, welche Ortsbestimmung den Bezugsrahmen für das bildet, was über den Weg eines Dinges gesagt wird. Um terminologische Mißverständnisse zu vermeiden, muß ich nun einige Worte zum theoretischen Rahmen sagen, innerhalb dessen ich mich bewege.

2. Syntaxregeln

Die Syntax, die ich meinen Überlegungen zugrundelege, ist so primitiv, wie sie für die hier verfolgten Zwecke noch eben sein kann. Ich tue so, als ließen sich die Ausdrücke des Deutschen mithilfe einer kategorialen Syntax beschreiben. Die syntaktische Beschreibung des Satzes (1) kann unter dieser Annahme etwa folgendermaßen aussehen.

(1a)

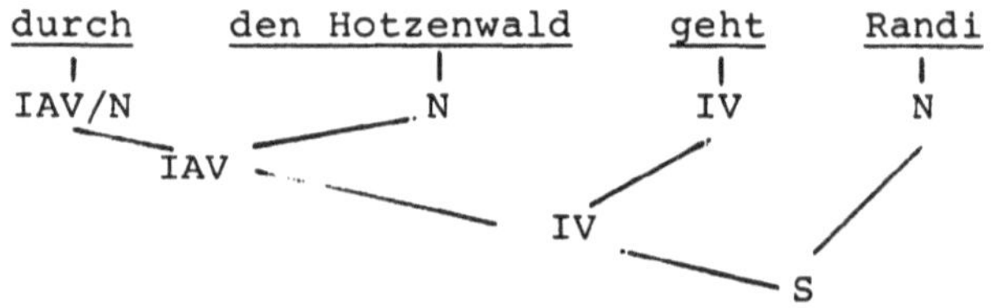

Dabei steht: "S" für "Satz" und "N" für "Name".
S und N werden syntaktische Grundtypen unserer Syntax genannt. Daneben gibt es noch zusammengesetzte Typen, die nach der folgenden Regel gebildet werden:

2.1 Definition: Wenn $\sigma, \tau_1, \ldots, \tau_n$ syntaktische Typen sind, dann ist auch $(\sigma/\tau_1, \ldots \tau_n)$ ein syntaktischer Typ unserer Syntax.

Beispiele für zusammengesetzte Typen:
"IV" ist eine Abkürzung für "S/N". Dies läßt sich lesen als "macht aus einem N ein S". "IV" erinnert natürlich an intransitives Verb. Typen der Form $\sigma/\tau_1, \ldots, \tau_n$ (Außenklammern werden weggelassen, wenn nicht benötigt) heißen Funktorentypen. Ausdrücke eines solchen Typs Funktoren.
"IAV" ist eine Abkürzung für "IV/IV", also für "(S/N)/(S/N)". Ein IAV ist also ein Ausdruck, der aus einem intransitiven Verb wieder eines macht. "IAV" erinnert an "Intransitivadverb" (im Gegensatz zu "Satzadverb").

IAV/N ist der Typ der Präpositionen. Eine Präposition macht aus einem Namen ein Intransitivadverb.

Die Ausdrücke einer kategorialen Sprache werden auf die denkbar einfachste Weise gebildet. Man hat ein (endliches) Lexikon, das den Wörtern der Sprache syntaktische Typen zuweist. Das hier vorausgesetzte Lexikon kann man dem Baum (1a) direkt ablesen. Jeder präterminale Knoten enthält als Etikett jeweils den Typ des Wortes, das am Endknoten klebt. Die zusammengesetzten Ausdrücke der Sprache bestimmen sich nach der folgenden Regel:

2.2 Definition. Wenn δ ein Ausdruck des Typs $\sigma/\tau_1,\ldots,\tau_n$ und $\alpha_1,\ldots,\alpha_n$ beziehungsweise Ausdrücke der Typen $\tau_1,\ldots,\tau_n$ sind, dann ist $(\delta, \alpha_1,\ldots, \alpha_n)$ ein Ausdruck des Typs σ.

Der syntaktische Rahmen, innerhalb dessen sich unsere Überlegungen bewegen, ist damit vollständig beschrieben.

3. Semantische Begriffsbildungen

Im nächsten Abschnitt werde ich noch einige semantische Begriffe benötigen, die ich hier ohne viel Kommentar vorstelle.

3.1 Definitionen. (a) Ein Tripel $\langle w,z,o\rangle$ bestehend aus einer Welt w, einem Zeitintervall z und einer Ortsbestimmung für z, o, nennen wir Situation (Kontext, Umstand). (b) Die Situation c_1 ist eine Teilsituation der Situation c_2 gdw. $c_1 = \langle w_1,z_1,o_1\rangle$, $c_2 = \langle w_2,z_2,o_2\rangle$ mit $w_1 = w_2$, $z_1 \subseteq z_2$ und $o_1(m) = o_2(m)$ für jedes $m \in z_1$.

Im folgenden nennen wir die Menge aller Ortsbestimmungen O. Die Menge aller Situationen, d.h. WxZxO, nennen wir C. Wir können nun sagen, was Propositionen sind.

3.2 Definition. (a) Eine Proposition ist eine Funktion aus C in die Menge der Wahrheitswerte {0,1}. (D.h., eine Proposition ordnet bestimmten Umständen Wahrheitswerte zu.) (b) Sei c eine Situation, p eine Proposition. Dann gilt: p ist wahr in c gdw. $p(c) = 1$. p ist falsch in c gdw. $p(c) = 0$. (c) Die Proposition q folgt (logisch) aus der Proposition p gdw. für jedes $c \in C$ gilt: wenn $p(c) = 1$, so $q(c) = 1$.

Ich benötige nun noch den Begriff der Eigenschaft.

3.3 Definition. (a) Eine n-stellige Eigenschaft ist eine Funktion aus A^n in die Menge der Propositionen. (A ist die Menge der möglichen Dinge!) (b) Sei ω eine n-stellige Eigenschaft, $\langle a_1,\ldots,a_n\rangle$ irgend ein n-Tupel von Individuen, c eine Situation. Dann gilt: ω trifft in c auf $\langle a_1,\ldots,a_n\rangle$ zu gdw. die Proposition $\omega(a_1,\ldots,a_n)$ in c wahr ist. ω trifft in c auf $\langle a_1,\ldots,a_n\rangle$ nicht zu gdw. $\omega(a_1,\ldots,a_n)(c) = 0$.

Einstellige Eigenschaften nenne ich im folgenden einfach "Eigenschaften". Ein Beispiel dafür, daß das Zutreffen einer Eigenschaft von einer Ortsbestimmung abhängen kann, lesen wir bei Schnitzler:

(4) Dieser aber sprach: "... Euer Freund aber, Tobias Klenk, wo immer ich es sein mag, hier in der Amtsstube bin ich es nicht."

Zur Erläuterung dessen, was da gemeint ist: Im Wirtshaus trifft es auf den Richter zu, daß er der Freund des Tobias Klenk ist, nicht aber vor Gericht in der Amtsstube.

4. Inhaltsregeln für "Randi" und "geht"

Von den verschiedenen Aspekten der "Bedeutung" eines Satzes interessiert mich hier nur derjenige, der für eine beliebige Äußerungssituation dieses Satzes festlegt, was mit ihm in dieser Situation ausgedrückt wird, falls damit überhaupt etwas ausgedrückt wird. Diesen Aspekt rekonstruiere ich als ein System von Verwendungsregeln, Inhaltsregeln genannt. Für jedes Wort gibt es eine Regel, die bestimmt, was mit ihm im Zusammenspiel mit anderen Wörtern in einer Situation ausgedrückt wird. Auf die Satzebene bezogen beschreiben diese Regeln, wie wir Propositionen ausdrücken. Ich beschreibe nicht, was mit diesen Propositionen gemacht werden kann, ob sie behauptet, in Frage gestellt oder vermutet werden - und wie man das macht. Insofern paßt dieser Ansatz genau zu den üblichen semantischen Theorien der möglichen-Welten-Semantiker. Der Typ von Regeln, die ich jetzt benutze, ist von Angelika Kratzer entwickelt worden.[2]

Ich möchte nun zunächst die Inhaltsregeln für den folgenden einfachen Satz angeben:

(5) Randi geht

Nach unserer Syntax wird er syntaktisch analysiert als

(5a) (geht, Randi).

Um zu beschreiben, welche Propositionen mit (5a) ausgedrückt werden können, müßten wir zuerst etwas über die Verwendung eines Namens wie Randi sagen. Das möchte ich aber hier nicht tun, weil dies ein kontroverses Gebiet ist, das mich hier nicht interessiert (Ich sympathisiere mit der Kripke-Donnellan-Kaplanschen Namenstheorie.)[3] Für die Zwecke dieses Beitrages genügt die folgende Annahme:

4.1 Wenn der Name Randi in der Situation <w,z,o> geäußert wird, dann wird damit entweder nichts bezeichnet oder genau eine weibliche Person, die in w zu irgendeiner Zeit existiert.

Wie man es macht, mit der Äußerung eines Namens etwas zu bezeichnen, das lasse ich völlig offen. Wichtig ist, daß der Inhalt (= das Bezeichnete) eines Namens kein Begriff, sondern ein Individuum ist.

Nun zur Inhaltsregel für geht. Dieses IV drückt in jeder Situation die Eigenschaft 'jetzt zu laufen' aus. In unserer Terminologie wird das so formuliert.

4.2 Sei c eine beliebige Situation, in der geht geäußert wird. Dann wird mit geht in c nur dann etwas ausgedrückt, wenn gilt:
(i) Es gibt genau eine Zeit z, die in der Situation c als gegenwärtige Zeit gilt und eine Ortsbestimmung für z, die in c als Bezugortsbestimmung gilt. Sei c eine Situation, welche die Bedingung (i) erfüllt, z_0 die Zeit, die in c als gegenwärtige zählt, o_0 die Ortsbestimmung, die in c als Bezugortsbestimmung zählt. Dann gilt: Mit der Äußerung von geht in der Situation c wird diejenige Eigenschaft ω ausgedrückt, die die Bedingungen (ii) und (iii) erfüllt:
(ii) ω ist nur für belebte Wesen erklärt;
(iii) Sei a irgendein belebtes Wesen. Dann ist $\omega(a)$ diejenige Proposition p, die nur für solche $\langle w,z,o\rangle$ erklärt ist, so daß a sich in w während der Zeit z_0 am Ort O_0 befindet. Für jedes solche $\langle w,z,o\rangle$ ist p wahr genau dann, wenn a in w während der Zeit z_0 am Ort o_0 läuft.

Ich nehme für das folgende an, daß jedes IV in einer beliebigen Äußerungssituation, in der es klar ist, welche Zeit als gegenwärtige Zeit gilt, stets eine Eigenschaft ausdrückt. Wir müssen nun noch erklären, wie man den Inhalt, der durch ein IV ausgedrückt wird, mit dem Inhalt, der durch ein N bezeichnet wird, zu einem S-Inhalt zusammenfügt. Dies leistet die folgende Regel.

4.3 Sei δ_{IV} ein IV, β_N ein N. Dann ist nach 2.2 $\alpha = (\delta_{IV}, \beta_N)$ ein S. Sei c eine beliebige Äußerungssituation von α, und seien c_1 und c_2 die Teilsituationen von c, in denen beziehungsweise δ_{IV} und β_N geäußert werden. [4] Dann wird mit α in c genau dann eine Proposition ausgedrückt, wenn gilt:
(i) Mit β_N wird in c_2 ein Ding a bezeichnet und δ_{IV} drückt in c_1 eine Eigenschaft aus, die für a erklärt ist.

Sei c eine Äußerungssituation von α, die diese Bedingung erfüllt. Sei <u>a</u> das durch β_N in c_2 bezeichnete Ding und ω die durch δ_{IV} in c_1 ausgedrückte Eigenschaft. Dann gilt:

(ii) α drückt in c die Proposition $\omega(a)$ aus.

Nun haben wir beschrieben, welche Inhalte wir mit (5a) ausdrücken können. Angenommen, ich äußere (5a) am 1. Januar 1978 während der Zeit von 0 Uhr bis 0 Uhr 3 Sekunden in der Casa Musette in Brione/sM. Angenommen, ich bezeichne mit *Randi* in der entsprechenden Teilsituation (d.h. von 0 Uhr 2 Sek.bis 0 Uhr 3 Sek.) meine Tochter, die Zeit so um Mitternacht herum sei die gegenwärtige, das Innere der Casa Musette der Bezugsort. Nach 4.2 drückt dann *geht* die Eigenschaft aus, so um Mitternacht herum in der Casa Musette zu gehen. Nach 4.3 drücke ich in dieser Situation die Proposition p aus, die nur für solche $\langle w,z,o\rangle$ erklärt ist, so daß meine Tochter sich in w um 0 Uhr des 1.1.1978 herum in der Casa Musette befindet (und deshalb in w während dieser Zeit natürlich auch existiert!). Für jedes solches $\langle w,z,o\rangle$ ist p wahr gdw. meine Tochter um 0 Uhr des 1.1.78 herum in w in der Casa Musette geht. Mit anderen Worten: ich drücke in der genannten Situation mit (5a) die Proposition aus, daß meine Tochter um die Mitternacht des 1.1.78 herum in der Casa Musette geht.

5. Die Inhaltsregel für "durch"

Wir kommen nun endlich zu *durch*. Ich habe in Abschnitt 1 bereits gesagt, daß wir Bewegungen relativieren müssen. Dem dient die folgende Definition.

5.1 <u>Definition</u>. Sei $c = \langle w,z,o\rangle$ eine beliebige Situation, a ein Ding, das sich in w während z am Ort o befindet. Dann gilt:

(a) <u>Die Bahn von a relativ zu c</u>, $B_{a/c}$ ist diejenige Funktion f, so daß für ein beliebiges $m \in z$ gilt:

$f(m) = \langle M_m, n_m\rangle$ mit:

M_m = { $x \in R$: Es gibt ein $y \in B_{a,w,z}(m)$, so daß x = y - dem Nullpunkt von o(m) }

n_m = der Nullpunkt von $B_{a,w,z}(m)$ - dem Nullpunkt von o(m). [5]

(b) <u>Der Weg von a relativ zu c</u>, $P_{a/c}$ ist die Funktion g, so daß für jedes $m \in z$ gilt: g(m) ist der Nullpunkt von $B_{a/c}(m)$.

Diese Definitionen bewirken anschaulich, daß die Bewegung des Ortes o annulliert wird. Der Weg $P_{a/c}$ hat am absoluten Ort genau die Form des Weges, den a in w während z <u>am</u> Ort o beschreibt.

Ich habe leider keinen Platz, um das an einem Beispiel vorzuführen. Wir können nun die Inhaltsregel für durch angeben.

5.2 Der Satz $\alpha = (((durch, \beta_N), \delta_{IV}), \gamma_N)$ werde in der Situation c geäußert. c_1, c_2, c_3 seien die Teilsituationen von c in denen beziehungsweise β_N, δ_{IV} und γ_N geäußert werden. Dann wird mit α in c genau dann eine Proposition ausgedrückt, wenn (i) und (ii) gelten:

(i) β_N und γ_N bezeichnen in c_1 und c_3 etwas. δ_{IV} drückt in c_2 eine Eigenschaft aus. Sei c eine Äußerungssituation von α, welche diese Bedingung erfüllt. Seien a und b die durch β_N und γ_N in c_1 bzw. c_3 bezeichneten Dinge, sei ω die durch γ_{IV} in c_2 ausgedrückte Eigenschaft, z_O die Zeit, die in c als gegenwärtige und o_O die Ortsbestimmung, die in c als Bezugsortsbestimmung zählt. (Vgl. dazu 4.2!) Dann drückt α in c diejenige Proposition q aus, so daß gilt:

(ii) q ist nur für solche $\langle w,z,o \rangle$ erklärt, so daß a und b sich in w während z_O am Ort o_O befinden. Für jedes solche $\langle w,z,o \rangle$ ist p wahr genau dann, wenn der Weg $P_{b/\langle w,z_O,o_O \rangle}$ führt durch das Gebiet von $B_{a/\langle w,z_O,o_O \rangle}$ (d.h. der Weg von b in w während z relativ zur Situation $\langle w,z_O,o_O \rangle$ führt durch das Gebiet, das a in w während z_O relativ zur genannten Situation okkupiert.)

Angenommen nun, ich äußere unseren Satz (2a) in einer Situation, für die die Bezugsortsbestimmung das Innere des Speisewagens des Intercityzuges "Räuber Hotzenplotz" ist. Dann drücke ich damit nach 5.2 eine Proposition aus, die nur für solche Umstände Sinn macht, unter denen sowohl Randi als auch der Hotzenwald im genannten Speisewagen sind. (Er muß da hineingekommen sein, wie weiland der Wald von Dunsinan in das Schloß des Macbeth.) Für solche Umstände ist die genannte Proposition dann wahr, wenn Randi im Speisewagen geht und sich dort durch den Hotzenwald bewegt. Der Einwand, den wir gegen Cresswells Wahrheitsbedingungen gebracht haben, ist hier nicht mehr möglich.

Es ist nun kein Problem mehr, auch Lokaladverbiale wie im Speisewagen zu behandeln. Die Funktion eines solchen Adverbials ist es, einen vorhandenen Bezugsort einzuschränken. Die Techniken dafür sind in Kratzer 1977 voll entwickelt. Ich hätte das Zusammenspiel von im und im Kreis herum um wie es im Satz (3) vorliegt, gerne noch diskutiert, aber: die mir zugestandenen Seiten sind voll.

ANMERKUNGEN

+ Ich möchte Jan Lerner für die intensive Diskussion danken,die ich mit ihm über diese Problematik führen durfte. Für Korrekturen danke ich ferner Pirmin Stekeler.

1 Die Menge Z der Zeitintervalle ist natürlich konstruiert aus einer Menge T von Zeitpunkten. T ist geeignet geordnet.Diese Ordnung überträgt sich dann teilweise auf Z. Ich kann diese Begriffe hier nicht motivieren, sondern verweise auf die Literatur, z.B. Cresswell 1973 und 76.

2 Vgl. Kratzer 1977.

3 Für die in diesem Zusammenhang einschlägigen Probleme siehe etwa Kaplan 1977.

4 Wenn ein Ausdruck α während der Situation $\langle w,z,o \rangle$ geäußert wird, heißt daß u.a., daß die Äußerung von α während z am Ort o stattfindet. Da eine individuelle Äußerung ein Individuum ist, findet hier Def. 1.2(f) Anwendung.

5 Für beliebige $x = \langle a_1,b_1,c_1 \rangle$, $y = \langle a_2,b_2,c_2 \rangle \in \mathbb{R}^3$ gilt natürlich: $x - y := \langle a_1-a_2, b_1-b_2, c_1-c_2 \rangle$.

LITERATUR

BENNETT, D.C.(1975): Spatial and Temporal Uses of English Prepositions. An Essay in Stratificational Semantics. London:Longman.

CRESSWELL, M.J.(1973): Logics and Languages. London: Methuen.

---(1976): Interval Semantics and Logical Words. Manuscript.

---(1977): Prepositions and Points of View. Manuscript. Erscheint in Linguistics and Philosophy.

KAPLAN, D.(1977): Demonstratives. An Essay on the Semantics, Logics, Metaphysics and Epistemology of Demonstratives and other Indexicals. Manuscript, März 1977.

KRATZER, Angelika / v.STECHOW, Arnim (1977): Äußerungssituation und Bedeutung. Literaturwissenschaft und Linguistik.Heft 23/24.

KRATZER, Angelika(1977): Kontexttheorie. Arbeitspapiere des SFB 99. Konstanz.

LERNER, Jan Y. (1973): Han går in i skogen - Semantische und syntaktische Analyse. The Nordic Languages and Modern Linguistics. 2. Stockholm.

LUTZEIER, P.(1974): Der 'Aspekt' Welt als Einstieg zu einem nützlichen Kontextbegriff für eine natürliche Sprache. Dissertation Stuttgart.

PIETSCH, R.(1977): Syntax und Semantik von Lokaladverbien in einer λ-Kategorialsprache. Staatsarbeit. Universität Freiburg.

v.STECHOW, Arnim (1977): Adverbien und Individuen. Originalpapier für Pavia.

TEMPUS, ADVERB, TEMPORALE FRAGE

Rainer Bäuerle

Bei der Untersuchung des Frage-Antwort-Dialogs hat es sich als fruchtbar erwiesen, den illokutionären Aspekt vom propositionalen Gehalt zu unterscheiden. Der propositionale Gehalt jeder Sequenz Frage-Antwort, so wird in EGLI(1974:119) angenommen, muß logisch einer Aussage äquivalent sein.

Beschäftigt man sich nun von dieser Ausgangsposition her mit der temporalen, d.h. der WANN-Frage, so stellt sich sofort das Problem der Interpretation der Tempora. Die Frage

(1) Wann eroberte Caesar Britannien?

ist nämlich deutlich eine Frage nach einem vergangenen Zeitpunkt, d.h. wir werden als mögliche Antworten Adverbien zulassen wie <u>gestern</u> oder <u>vor ein paar hundert Jahren</u>, nicht aber solche wie <u>morgen</u> oder <u>in ein paar hundert Jahren</u>. Diese Einschränkung der Antwortmöglichkeiten ist natürlich auf das Tempus im Fragesatz zurückzuführen, so daß wir in etwa die folgende Repräsentation des Fragedialogs mit einem tiefenstrukturellen Tempusmarker annehmen müssen:

(2) λx_{Adv}(((erobern, Caes, Brit), Past), x_{Adv}) - (Gestern).

Nun haben wir aber angenommen, daß ein minimaler Dialog der Form $((\lambda x_{Adv}\ S)Adv)_{Dial}$ im propositionalen Gehalt äquivalent ist mit jenem S', das aus S durch Substitution von Adv für jedes durch den Lambda-Operator abgebundene Vorkommen von x_{Adv} gewonnen wurde. Somit ist also (2) äquivalent mit

(3) (((erobern, Caesar, Britannien), Past), gestern)

als Repräsentation der Aussage

(4) Gestern eroberte Caesar Britannien.

Das Problem ist nun, daß unsere Überlegungen zur Analyse des Fragedialogs zur Annahme eines tiefenstrukturellen Tempus in (3) führten, während nach gängiger Ansicht (3) eben nicht die tiefenstrukturelle Repräsentation von (4) sein kann. Nach der Argumentation von CRESSWELL(1973:195f) kann (3) höchstens

(5) Gestern hatte Caesar Britannien (bereits) erobert
wiedergeben, da es paraphrasierbar ist als

(6) Es war zu einer Zeit vor gestern der Fall, daß Caesar Britannien erobert.

Aus ganz analogen Gründen muß Cresswell auch

(7) (((erobern, Caesar, Britannien), gestern), Past)

als Tiefenstruktur für (4) verwerfen, denn dies liest sich als

(8) Es war zu einer vergangenen Zeit der Fall, daß am Tag zuvor Caesar Britannien erobert.

So kommt Cresswell zu demselben Schluß wie schon vor ihm WUNDERLICH(1970:166): Tempus und sprechzeitrelatives Adverb treten nicht zusammen, sondern nur alternativ in der Tiefenstruktur auf. Damit wäre die (4) zugrundeliegende Struktur aber

(9) ((erobern, Caesar, Britannien), gestern).

Nun, so vermerkt bereits Wunderlich (p.181), gilt dies aber nicht für sprechzeitüberlappende Adverbien wie <u>heute</u>: hier kann nur das Tempus festlegen, ob auf Vergangenes oder Zukünftiges referiert wird. Und für eine Analyse des propositionalen Gehaltes der WANN-Frage ist, wie wir gesehen haben, dieser Ansatz ebenfalls unbrauchbar. Unsere Aufgabe ist es deshalb, wegzukommen von einer Analyse, bei der sich Tempus und Adverb gegenseitig ausschließen, und hinzuführen zu einer Erklärung des Zusammenwirkens von Tempus und Adverb, die ja ganz natürlich miteinander vorkommen.

Cresswells Analyse, die zur Ablehnung eines tiefenstrukturellen Vergangenheitsoperators führt, beruht nun auf einer Semantik der Tempora, die man die INDEFINITE nennen kann, und die ganz besonders der temporalen Logik zugrunde liegt:

(10) $Pp =_{def} (\exists t)\ (Utn\ \&\ R_t(p))$

(In dieser Definition von RESCHER/URQUHART(1971:52) sind p und t Variable für Propositionen beziehungsweise Zeitpunkte, n steht für <u>now</u>, U für die Vorgängerrelation, P ist der Vergangenheitsoperator, und $R_t(p)$ heißt <u>p is realized at t</u>.)

Diese indefinite Interpretation der Tempora, angewandt auf natürliche Sprachen, ist auch dann problematisch, wenn wir einmal von dem Problem des Zusammenwirkens mit den Adverbien absehen.

Sei nämlich

(11) (∃t) (Utn & R_t(kommen, Hans))

jetzt wahr, so muß es auch von nun an immer wahr sein, denn t liegt ja auch vor jedem Zeitpunkt nach n. Damit kann (11) aber keine tiefenstrukturelle Repräsentation für

(12) Hans kam

sein, denn (12) muß nicht von nun an immer wahr sein, sondern kann auch in Zukunft zu einer Lüge verwendet werden. Und zur Verneinung eines Konstrukts wie (11) bemerkt HALL-PARTEE(1973):

> The deictic use of the past tense morpheme appears in a sentence like ... 'I didn't turn off the stove'. When uttered, for instance, halfway down the turnpike, such a sentence clearly does not mean either that there exists some time in the past at which I did not turn off the stove or that there exists no time in the past at which I turned off the stove. The sentence clearly refers to a particular time ... whose identity is generally clear from the extralinguistic context ...

Der INDEFINITEN Interpretation der Tempora, die einen in den natürlichen Sprachen eher marginalen Gebrauch repräsentiert (der dann kontextuell oft deutlich markiert ist), zum Beispiel das Englische

(13) Have you ever been to London?

oder das Deutsche

(14) Warst du schon einmal in London?

wird bei Hall-Partee die Forderung nach einer DEFINITEN Interpretation gegenübergestellt. Zu Recht wird damit eine über zweihundert Jahre alte Unterscheidung neu belebt. So lesen wir in BEAUZÉE(1767:450-51):

> Quant aux propositions d'une vérité contingente, elles ont nécessairement des rapports différents aux diverses époques; ... par conséquent on ne peut y employer que des Temps définis; ou si l'on y emploie les indéfinis, ce ne peut être que dans un sens défini & avec application à des époques déterminées.

Interessant ist hier, wohlgemerkt, nicht die Einteilung der Tempora in definite und indefinite, sondern der Vermerk des definiten Gebrauchs, die 'application à des époques déterminées'. Auch daß ein solcher definiter Gebrauch kontextabhängig ist, war Beauzée bekannt (p.439):

> ... le présent indéfini ne se rapporte lui-même à aucune époque déterminée, ce sont les circonstances du discours qui déterminent celle à laquelle on doit le rapporter en chaque occasion.

Was Beauzée die 'circonstances du discours' nennt, ist bei Hall-Partee der extralinguistische Kontext, und ist bei unserem Problem des Zusammenwirkens von Tempus und Adverb der linguistische Kontext oder Kotext. Es handelt sich also um ein und dasselbe Problem, denn die in der Sprechsituation mitverstandene 'époque déterminée', an anderer Stelle auch 'époque de comparaison' genannt, läßt sich natürlich auch immer sprachlich spezifizieren: wenn A den B zu t frägt, warum er nicht gekommen sei, so meint er damit, zum Beispiel in der Sprechsituation C, warum B am Abend vor t nicht, wie verabredet, ins Kino kam.

Nun ist natürlich auch die indefinite Interpretation von der Sprechsituation abhängig, es bedarf daher einer Präzisierung. Beauzée unterscheidet 'l'instant même de l'acte de la parole' und 'l'époque de comparaison'. Wir verwenden hier die Begriffe SPRECHZEIT und BETRACHTZEIT. Der indefinite Ansatz berücksichtigt nur die kontextuelle Information der Sprechzeit, während der definite auf Sprechzeit und Betrachtzeit basiert:

> ... the most obvious method of elimination is to attach an index to the statement indicating at what 'position' in the order of events the event or events described by the statement occured, TOGETHER with an index attached to the statement indicating at what 'position' in the order of events the statement was made. (HARRÉ, 1960)

Dabei ist die Sprechzeit nur ein Spezialfall des allgemeineren Begriffs der EVALUATIONSZEIT. So bezeichnet die Adverbiale vor zwei Wochen eine Zeit zwei Wochen vor der Evaluationszeit, die nun mit der Sprechzeit identisch sein kann, so in

(15) Vor zwei Wochen kam Hans,

dies aber nicht sein muß, wie das nächste Beispiel zeigt:

(16) Gestern vor zwei Wochen kam Hans.

Dagegen bezeichnet gestern den Tag vor dem Tag, der die Evaluationszeit enthält, wobei diese in jedem Fall mit der Sprechzeit identisch ist.

Damit sind wir nun in der Lage, die folgenden Unterscheidungen zu treffen: eine Datumsadverbiale spezifiziert die Betrachtzeit ohne Rekurs auf den Kontext (Hans kam am 1.9.1977); eine evaluationszeitrelative Adverbiale (mit dem Spezialfall der sprechzeitrelativen Adverbiale) spezifiziert die Betrachtzeit in Abhängigkeit von einer gegebenen Evaluationszeit (Beispiele (15) und (16)); die Tempora in definiter Auffassung, so ist man zunächst geneigt anzunehmen, geben die zwischen der Evaluationszeit und der Betrachtzeit bestehende Relation der Vor-, Gleich- oder Nachzeitigkeit an.

Diese Auffassung würde jedoch in ähnliche Schwierigkeiten führen wie der indefinite Ansatz: das Imperfekt wäre nun zwar mit einem Adverb wie gestern kompatibel, nicht aber mit heute, denn das von diesem Adverb spezifizierte Zeitintervall liegt nicht vor der Evaluationszeit, sondern überlappt sie. Deshalb müssen wir annehmen, daß das Tempus aus einer vorgegebenen Betrachtzeit eine Betrachtzeit' in Abhängigkeit von der Evaluationszeit herausfiltert. Für das Imperfekt heißt das konkret, daß die Betrachtzeit' der vor der Evaluationszeit liegende Teil der Betrachtzeit ist. Für den Fall einer Betrachtzeit heute ist dies ein echtes Teilintervall von heute, für eine Betrachtzeit gestern ist dies dasjenige Teilintervall von gestern, das mit gestern identisch ist. Auch auf das Futur läßt sich diese Analyse leicht übertragen.

Ein spezieller Präsens-Operator wird nun in der Temporallogik gemeinhin als überflüssig betrachtet:

> Note however that construing the tense operators as statement operators requires the verb in the (temporally indefinite) interior statement to have some tense or other. ... not only does the statement operator T signify present tense but the interior statement is in the present tense as well. Thus the T operator is superfluous and can be dispensed with. ... The function of F and P is to strip all but present tenseness from the main verb in the original statement and pack the tense into the prefix.
> (MC ARTHUR,1976:3)

Wenn nun aber wie hier das Verb im Skopus der Operatoren PAST, FUTURE, aber auch YESTERDAY und TOMORROW, als präsentisch angesehen wird, dann ist klar, daß 'present tenseness' nur eines heißen kann: Übereinstimmung mit der Betrachtzeit. Dem Präsens-

Operator käme also nur die Deutung als Identitätsfunktion zu, und in diesem Sinn ist er tatsächlich überflüssig. Auch diese Deutung des Präsens findet sich bereits bei Beauzée (p.433):

> Un présent défini est donc actuel, antérieur ou postérieur, selon qu'il exprime la simultanéité d'existence à l'égard d'une époque déterminément actuelle, antérieure, ou postérieure.

Diese These belegt Beauzée dann im Anschluß durch Beispiele für aktuelles Präsens, für praesens historicum und für praesens pro futuro.

Aus dieser Diskussion ergibt sich nun, daß das, was Cresswell mit einer Tiefenstruktur wie

(17) ((John, sleep), yesterday)

erfaßt, eben nicht

(18) John slept yesterday

ist, sondern ein historisches Präsens.

Daß das Präsens in der sprachwissenschaftlichen Literatur überwiegend als 'present tenseness' im Hinblick auf die Sprechzeit gedeutet wird, ist nicht weiter verwunderlich, da dies ja im Rahmen einer indefiniten Interpretation der Tempora geschieht die eben lediglich auf die Sprechzeit bezogen ist. Dies bedingt aber, daß praesens historicum und praesens pro futuro als Ausnahmen dargestellt und behandelt werden müssen, daß die Kompatibilität des Präsens mit allen möglichen Adverbien, auch wenn diese eine von der Sprechzeit sehr weit entfernte Betrachtzeit spezifizieren, keine befriedigende Erklärung findet.

In einem definiten Ansatz finden diese Phänomene nun eine durchaus Regel-mäßige Erklärung: 'present tenseness' wird im Hinblick auf die Betrachtzeit gedeutet, diese aber mag in der Zukunft liegen (<u>Morgen kommt Hans</u>), sie mag in der Vergangenheit liegen (<u>55 v.Chr. erobert Caesar Britannien</u>), oder aber sie mag einen Gegenwartsbezug ausdrücken (<u>Hans kommt</u>).

Die Erklärung dafür, daß wir nicht alle diese Strukturen als gleichberechtigt empfinden, daß uns ein praesens pro futuro vielleicht natürlicher vorkommt als ein praesens historicum, daß das letztere an spezielle Kontexte gebunden sein mag, dies scheint mir eher in das Gebiet der Sprachverwendung zu gehören.

Neben den bisher eingeführten Begriffen der Sprechzeit, der Evaluationszeit und der Betrachtzeit muß nun noch der Begriff der AKTZEIT thematisiert werden. Dies wird vielleicht klar an einem Zitat aus GUENTHNER(1977), wo ebenfalls die Wichtigkeit der Betrachtzeit unterstrichen wird. Guenthner gibt nämlich die Deutung des Vergangenheitsoperators P wie folgt an:

(19) $[P\emptyset]^{I}_{t,g} = 1$ iff $[\emptyset]^{I}_{t'} = 1$ for some $t' \in I$ and $t'<t$

Das heißt, daß $P\emptyset$ wahr ist zur Zeit t in Bezug auf das Referenzintervall I (= Betrachtzeit) gdw. $\emptyset$ zu einer Zeit t' aus I wahr ist, die vor t liegt. Guenthner quantifiziert in dieser Deutung über die Aktzeit t', dies gehört aber nach meiner Meinung nicht zur Semantik des Tempusoperators, denn nur aus diesem Grund handelt sich Guenthner die folgende Schwierigkeit ein:

> We should add that there are some problems with negation in such contexts. For instance, a sentence like <u>Yesterday John didn't play tennis</u> cannot be translated into YP¬Tj - which might be true even if John did play at some but not at all times - but into Y¬PTj. (GUENTHNER,1977:31)

Nach unseren bisherigen Ausführungen spezifiziert das Tempus das Betrachtzeitintervall', nichts weiter. Innerhalb dieser Betrachtzeit' muß das betreffende Ereignis nun nicht unbedingt <u>at some t</u>' eingetreten sein, es kann einmal, zweimal, öfter oder eben überhaupt nicht eingetreten sein. Mit anderen Worten: Quantifikation über der Aktzeit ist Aufgabe der Frequenzadverbien. Konsequenterweise muß nach dieser Auffassung jeder Satz ein Frequenzadverb enthalten, auch wo dieses morphologisch nicht realisiert ist, wie in Guenthners Beispiel <u>Yesterday John didn't play tennis</u>. Das eine morphologisch nicht realisierte Frequenzadverb ist das, das die von Guenthner in die Tempussemantik aufgenommene Existenzquantifizierung 'for some $t' \in I$' leistet. Es ergeben sich dabei keine Negationsprobleme, der Satz wird weder mit YP¬Tj noch mit Y¬PTj übersetzt, sondern mit YP¬Adv_fTj.

Nun wird in den zitierten Definitionen von Rescher bis Guenthner die Aktzeit immer als Zeitpunkt angegeben, deshalb sei hier noch ein Zitat aus GABBAY(1976:175) angeführt:

> We see that some verbs can be evaluated at a moment of time (e.g. win) while others need a period of time! This means that $D_t(p)$ should be replaced by $D_T(p)$, T is an interval.

Wenn ich für einen Satz wie Ich habe gestern mehrmals gesungen Zeitpunkte als Aktzeit nehme, dann wird dieser Satz auch dann richtig, wenn ich nur einmal gesungen habe, da dieses einmalige Singintervall sicher mehrere Zeitpunkte enthält. Es ist aber klar, daß hier auf mehrere Singereignisse referiert wird. Ebensowenig läßt sich bei Annahme einer punktuellen Aktzeit eine vernünftige Semantik für durative Adverbien angeben, wie sie für den folgenden Satz gebraucht wird:

(20) Ich habe gestern zwei Stunden lang gesungen.

Der Übergang von Aktzeitpunkten zu Aktzeitintervallen ist also ein notwendiger, wenn auch kein hinreichender Schritt. Denn auch ein Aktzeitintervall, kann man einwenden, enthält nicht nur mehrere Zeitpunkte, sondern auch mehrere Teilintervalle. Es ist dies ein deutlicher Hinweis darauf, daß es dringend erforderlich ist, diesen Ansatz durch ein entsprechendes Ereigniskonzept zu sichern und zu erweitern.

Nachdem nun die definite Tempusanalyse die Voraussetzung geschaffen hat für eine Analyse der temporalen Frage, muß noch einmal auf die möglichen Antworten eingegangen werden, denn nicht auf jede WANN-Frage kann mit jedem Temporaladverb geantwortet werden. Einschränkungen der Antwortmöglichkeiten ergeben sich nämlich durch die möglichen Kombinationen des Frageworts mit einem oder mehreren Adverbien im Fragesatz.

Ist das Fragewort mit keinem Adverb kombiniert, so wird der gesamte Adverbialkomplex erfragt, mögliche Antworten sind alle zulässigen Sequenzen von Temporaladverbien, sofern sie nur mit dem Tempus des Fragesatzes verträglich sind. Es ist ein anderes, in Abhängigkeit von der Fragesituation zu klärendes Problem, wann eine solche Antwort erschöpfend bzw. präzise genug ist.

Eine Adv_t-Frage kann nun zum Beispiel beantwortet werden mit

(21) gestern, heute, vor einer Woche ...

oder mit

(22) am Nachmittag, um 17 Uhr ...

Die unter (22) aufgeführten Adverbiale nennt WUNDERLICH(1970: 186f) Adv_k (k = kontextrelativ), die unter (21) aufgeführten umfassen Wunderlichs Kategorien Adv_s(s = sprechzeitrelativ) und $Adv_ü$ (ü = Sprechzeiteinschluß oder -unmittelbarkeit). Die Klasse der $Adv_ü$ wurde von Wunderlich vor allem deshalb von den Adv_s getrennt, weil nach seinem Ansatz bei $Adv_ü$ ein tiefenstrukturelles Tempus abgeleitet werden muß, bei Adv_s aber keines stehen darf. Nach der hier vorgeschlagenen definiten Interpretation ist eine solche Trennung aber überflüssig, so daß man mit Adv_s und Adv_k auskommen kann.

Als sprechende Begriffe scheinen mir die Termini 'sprechzeitrelativ' und 'kontextrelativ' freilich nicht ganz geglückt, denn unter die Adv_s fallen auch solche Adverbien, die lediglich evaluationszeitrelativ sind (vor einer Woche), während beide Gruppen, Adv_s und Adv_k, eine kontextabhängige Deutung haben. Nun ist die Evaluationszeit der Adv_k immer eine anderweitig spezifizierte oder mitverstandene Betrachtzeit, aus der ein Teilintervall herausgefiltert wird (man beachte die Parallele zu den Tempora), man könnte diese Gruppe also Adv_E (E = Betrachtzeit-Einschränkend) nennen. Die Adv_s andererseits spezifizieren alle möglichen Relationen zwischen Betrachtzeit und Evaluationszeit (vorher, nachher, überlappend), nur nicht die Inklusionsrelation $T_B \subseteq T_E$, nennen wir sie also Adv_S (S = Betrachtzeit-Spezifizierend; wie vage diese Spezifikation auch sein mag) mit den Untergruppen Adv_{Ss} (s = sprechzeitrelativ) und Adv_{Se} (e = evaluationszeitrelativ). Die lineare Abfolge ist dabei

$$Adv_{Ss} - Adv_{Se} - Adv_E.$$

Wir können nun unserer Typologie der temporalen Frage neben der Adv_t-Frage noch die Adv_S-Frage und die Adv_E-Frage hinzufügen. Eine Adv_{Ss}-Frage oder eine Adv_{Se}-Frage hingegen gibt es nicht. Die Kombination des Frageworts mit einem Adv_{Ss} oder einem Adv_{Se} ist immer eine Adv_E-Frage:

(23) Wann kam Hans gestern?

(24) Wann kam Hans vor einer Woche?

Beide Fragen können durch am Nachmittag, nicht aber durch vor einer Woche oder durch gestern beantwortet werden.

Bei der Kombination des Fragewortes mit einem Adv_E besteht eine oberflächenstrukturelle Ambiguität: es kann sich um eine Adv_E-Frage oder um eine Adv_S-Frage handeln.

(25) Wann kam Hans am Nachmittag? - Letzte Woche

(26) Wann kam Hans am Nachmittag? - Um 17 Uhr

Diese Ambiguität besteht nicht mehr, wenn noch ein Adv_S hinzukommt. Es kann sich dann wiederum nur um eine Adv_E-Frage handeln, denn nur in (26), nicht in (25), kann ein Adv_S eingefügt werden.

LITERATUR

BEAUZÉE, Nicolas (1767): Grammaire générale ou exposition raisonnée des éléments nécessaires du langage. I. Faksimileausgabe Stuttgart: Frommann-Holzboog (1974).

CRESSWELL, Max (1973): Logics and languages. London:Methuen.

EGLI, Urs (1974): Zur Integration der Semantik in die Grammatik. Kronberg: Scriptor.

GABBAY, Dov M. (1976): Investigations in modal and tense logics. Dordrecht: Reidel.

GUENTHNER, Franz (1977): Time schemes, tense logic and the analysis of English tenses. (vervielf.).

HALL-PARTEE, Barbara (1973): "Some structural analogies between tenses and pronouns in English". The Journal of Philosophy 70: 601-609.

HARRÉ, Rom (1960): "Linguistic foundations for a time logic". Atti del XII congresso internazionale di filosofia. 5: 213-220. Firenze: Sansoni.

McARTHUR, Robert P. (1976): Tense logic. Dordrecht: Reidel.

RESCHER, Nicholas / URQUHART, Alasdair (1971): Temporal logic. Wien: Springer.

WUNDERLICH, Dieter (1970): Tempus und Zeitreferenz im Deutschen. München: Hueber.

ZUM GEBRAUCH VON 'ALORS QUE', 'PENDANT QUE', 'TANDIS QUE', 'DES QUE' UND 'AUSSITOT QUE'

Hartmut Pott

Im folgenden werde ich mich mit den geläufigen Temporalsubjunktionen 'alors que', 'pendant que' und 'tandis que' einerseits und 'dès que' und 'aussitôt que' andererseits befassen. Ich verwende den Terminus 'Subjunktion' abkürzend für den Begriff 'Konjunktion der Subordination', der unhandlich ist und den ich für paradox halte. Statt 'Subjunktion' wird auch gelegentlich SUB-Element gesagt. Es soll nun versucht werden, Bedingungen syntaktischer und/oder semantischer Natur aufzuspüren und zu formulieren, die erfüllt sein müssen, damit die einzelnen Subjunktionen im Satz bzw. als Bindeglieder zwischen zwei Sätzen funktionieren können. Allen fünf zu untersuchenden SUB-Elementen ist gemeinsam, daß sie die Gleichzeitigkeit von wenigstens zwei Aussagen bzw. der diesen zugrundeliegenden Sachverhalte markieren.

(1) Alors que l'un disait oui, l'autre disait non.

(2) Pendant qu'il faut agir, on fait des discours.

(3) Tandis que sa famille s'épuise au travail, il fréquente les cafés.

(4) Dès que ne m'accapare plus le travail, j'appartiens à la tristesse.

(5) Aussitôt qu'il a travaillé, son frère a dormi.

Die Tiefenstrukturen zu (1) - (5) haben folgendes Aussehen:

(6) $[F(a,t_i)]^{\ast}(p),[G(b,t_j)]^{\ast}(q)$, wobei F = dire oui, G = dire non, a = l'un, b = l'autre

(7) $(\exists x)[F(x)]^{\ast}(p'),[O(p',t_i)]^{\ast}(p),[H((\iota y)[G(y)],t_j)]^{\ast}(q)$, wobei F = agir, O = ought (in der Modallogik übliches Symbol) für il faut, H = faire des discours, $(\iota y)[G(y)]$ = on, zu lesen als: alle die y, für die gilt: y ist ein G, G = homme

(8) $(\exists x)[F(x).R(a,x).G(x,t_i)]^{\ast}(p),(\exists y)[H(y).S(a,y,t_j)]^{\ast}(q)$, wobei F = famille, R = avoir, G = s'épuiser au travail, H = café, S = fréquenter, a = il (die Ableitung des Possessivadjektivs aus 'avoir' interessiert hier nicht weiter)

(9) $(\exists x)[F(x).R(x,a,t_i).\overline{R(x,a,t_j)}.(i<j)]^{\ast}(p),(\exists y)[G(y).S(a,y,t_k)]^{\ast}(q)$, wobei F = travail, R = accaparer, G = tristesse, S = appartenir à

(10) $[F(a,t_i)]^{*}(p),[G(b,t_j)]^{*}(q)$, wobei F = travailler, G = dormir, a = il, b = le frère de a
Für die q-Aussage ließe sich auch genauer schreiben: $(\exists x)[H(x).R(a,x).G(x,t_j)]^{*}(q)$, wobei H = frère, R = avoir, G = dormir (wie oben), jedoch ist diese Unterscheidung für unsere Zwecke hier ohne Belang.

In allen angegebenen Tiefenstrukturen (6) - (1o) finden sich Temporalangaben t_i, t_j, t_k, die ich als Temporalindividuen auffasse und die ich, um die Gleichzeitigkeit auszudrücken, in folgender Weise gleichsetzen kann:

(11) $t_i = t_j$ oder abgekürzt: $i = j$,

was beispielsweise für (6) ergibt:

(12) $[F(a,t_i)]^{*}(p),[G(b,t_j)]^{*}(q),(i=j)$

Vorzeitigkeit und Nachzeitigkeit wären dann folgerichtigerweise durch:

(13) $i < j$ bzw. $i > j$

auszudrücken, wie bereits in (9) angeklungen ist. Dabei gehe ich davon aus, daß die Zeit kontinuierlich abläuft und von einem beliebigen zurückliegenden Zeitpunkt an gemessen zu einem anderen Zeitpunkt eine gewisse Anzahl von Zeiteinheiten abläuft, deren Betrag immer größer wird, je später der zweite Zeitpunkt liegt.

Der Umstand, daß die SUB-Elemente 'alors que', 'pendant que', 'tandis que', 'dès que' und 'aussitôt que' in (1) - (5) austauschbar sind, ohne daß sich an der Akzeptabilität etwas ändert - als Beispiele mögen (14) - (18) dienen - bedeutet allerdings nicht, daß die genannten SUB-Elemente etwa äquivalent seien:

(14) Pendant que l'un disait oui, l'autre disait non.

(15) Tandis qu'il faut agir, on fait des discours.

(16) Dès que sa famille s'épuise au travail, il fréquente les cafés.

(17) Aussitôt que ne m'accapare plus le travail, j'appartiens à la tristesse.

(18) Alors qu'il a travaillé, son frère a dormi.

Als Beweis dafür dienen folgende Beispiele, die mit 'alors que', 'pendant que' und 'tandis que' akzeptiert werden, nicht jedoch mit 'dès que' und 'aussitôt que':

(19) Alors qu'il partait, nous arrivâmes.

(20) Pendant que Grandet raccomodait son escalier vermoulu, les trois Cruchot frappèrent à la porte.

(21) Tandis qu'il dormait, je me suis dérobé d'auprès de lui.

(22) *Dès qu'il partait, nous arrivâmes.

(23) *Aussitôt qu'il dormait, je me suis dérobé d'auprès de lui.

Umgekehrt sind die folgenden Sätze mit 'dès que' und 'aussitôt que' akzeptabel, während sie mit 'alors que', 'pendant que' und 'tandis que' abgelehnt werden:

(24) Dès qu'il fut parti, l'autre arriva.

(25) Aussitôt qu'arrivé, il se coucha.

(26) Dès que vous voudrez, restez.

(27) *Alors qu'il fut parti, l'autre arriva.

(28) *Pendant qu'arrivé, il se coucha.

(29) *Tandis que vous voudrez, restez.

Die Tatsache, daß bei Tempusangleichung Sätze wie (22) und (23) zu akzeptablen Ergebnissen führen:

(30) Aussitôt qu'il partait, nous arrivions.

(31) Dès que Grandet raccomodait son escalier vermoulu, les trois Cruchot frappaient à la porte.

(32) Aussitôt qu'il dormait, je me dérobais d'auprès de lui.,

könnte nun Anlaß sein zu meinen, die Nichtakzeptabilität von (22) und (23) rühre vom unterschiedlichen Tempusgebrauch in $(S)_p$ und in $(S)_q$ her, d. h. es müsse lediglich die Bedingung

$$\text{Tempus } (S)_p = \text{Tempus } (S)_q$$

eingeführt werden, um das Problem zu lösen; doch so einfach liegen die Dinge nicht, wie z. B. (24) - (26) zeigen. In (24) steht $(S)_p$ im passé antérieur, $(S)_q$ im passé simple; in (25) liegt eine Ellipse vor für: 'aussitôt qu'il fut arrivé, il se coucha' mit denselben Tempusverhältnissen wie in (24). In (26) steht $(S)_p$ im Futur, $(S)_q$ im Imperativ.

Dennoch bietet der Tempusansatz Erklärungsmöglichkeiten, die nicht unbedingt von der Hand zu weisen sind, wenn man sich nicht auf das Tempus des gesamten Verbalkomplexes bzw. der gesamten Aussage bezieht, sondern auf das Tempusmorphem des Auxiliarteils; so ist das Auxiliar in (24) und in (25) in der nichtelliptischen Form 'fut'. Schwierigkeiten macht dagegen (26). Wenn man aber davon ausgeht, daß der Imperativ seinem Wesen nach eigentlich eine Modifikation des Futurs ist, läßt sich diese Schwierigkeit umgehen. Die Bedingung wäre dann etwa folgendermaßen zu formuliern:

$\text{Tempus (AUX)}_{(S)_p} = \text{Tempus (AUX)}_{(S)_q}$, wobei der Imperativ als Modifikation des Futurums aufzufassen ist.

Voll befriedigen kann dieser Lösungsweg allerdings nicht, da das Gebot der Gleichzeitigkeit bei einer solchen rein morphosyntaktischen Betrachtungsweise auf der Strecke bleibt, denn in (24) und (25) ist die p-Aussage an sich vorzeitig zur q-Aussage. Die Lösung wird eher im Bereich der Semantik zu suchen sein, vor allem da die Ablehnung von (27) - (29) auf die eben versuchte Weise nicht in den Griff zu bekommen ist. Dazu ist zu bemerken, daß die Sätze:

(1) Alors que l'un disait oui, l'autre disait non.

und

(33) Dès que l'un disait oui, l'autre disait non.

bzw.

(34) Aussitôt que sa famille s'épuise au travail, il fréquente les cafés.

und

(3) Tandis que sa famille s'épuise au travail, il fréquente les cafés.

durchaus nicht bedeutungsgleich sind.

Tatsächlich ordnen die verschiedenen Grammatiken und Wörterbücher 'alors que', 'pendant que' und 'tandis que' einerseits und 'dès que' und 'aussitôt que' andererseits zwei verschiedenen Rubriken zu - eine Vereinfachung, wie wir später sehen werden. So werden z. B. bei KLEIN-STROHMEYER und bei BERTAUX-LEPOINTE - die Liste ließe sich beliebig erweitern - 'alors que'. 'pendant que' und 'tandis que' mit 'während' und modifizierenden Erläuterungen übersetzt, 'dès que' und 'aussitôt que' mit 'sobald als'.

Es wäre nun ein Irrtum anzunehmen, die SUB-Elemente selbst würden diese Differenzierung bewirken; vielmehr ist davon auszugehen, daß die Tiefenstrukturen, die die Verwendung der unterschiedlichen SUB-Elemente zulassen, in sich nicht eindeutig sind, so daß die unterschiedlichen SUB-Elemente lediglich den ambigen Charakter der Tiefenstruktur hervorkehren bzw. in verschiedener Weise interpretieren lassen. Denn, wenn es sich anders verhielte, so würde ein Funktionselement der Oberfläche (oder zumindest des oberflächennahen Bereichs die Bedeutung der Tiefenstrukturen verändern können, was per definitionem ausgeschlossen ist.

Halten wir also fest: die SUB-Elemente interpretieren mögliche Verhältnisse von Teilaussagen zueinander, und diese Verhältnisse sind semantischer Natur, und die Möglichkeiten zur unterschiedlichen Interpretation liegen im Sinn der Aussagen selbst.

ir müssen also, um die Bedingungen für den Gebrauch der unter-
chiedlichen SUB-Elemente formulieren zu können, die Tiefenstruk-
uren präziser fassen. Das bedeutet für das vorliegende Problem
ine Präzisierung des Begriffs der Gleichzeitigkeit. Dazu komme ich
uf die Beispiele (19) - (23) zurück, die ich noch um (35) und (36)
rweitern will:

(35) *{Aussitôt que / Dès que} Grandet raccomodait son escalier vermoulu, les trois Cruchot frappèrent à la porte.

(36) *{Dès qu' / Aussitôt qu'} il dormait, je me suis dérobé d'auprès de lui.

iesen Sätzen ist wie gesagt gemeinsam, daß $(S)_p$ im imparfait und $S)_q$ im passé simple bzw. passé composé stehen. Aber die Konsequenz araus ist weitergehend als die einfache Morphemverschiedenheit, die ir bereits behandelt haben.

Es ist allgemein bekannt, daß das Französische durch die Tempus-orpheme, insbesondere durch den Kontrast von imparfait vs. passé imple bzw. passé composé verschiedene Aspekte zum Ausdruck bringt, ie ohne diesen Kontrast nicht unbedingt gegeben sein müssen. Das eißt mit anderen Worten für unsere Beispiele, daß gerade durch den ontrast von imparfait in $(S)_p$ und passé simple bzw. passé composé n $(S)_q$ eindeutig gemacht wird, daß es sich bei $(t)_p$ um einen Zeit-aum und bei $(t)_q$ um einen Zeitpunkt handelt; und gerade diese larstellung, daß $(t)_p$ ein Zeitraum ist, steht im Widerspruch zur erwendung von 'dès que' und 'aussitôt que' im genannten Zusammen-ang. Die Bedingung für den Gebrauch dieser SUB-Elemente muß dann twa lauten:

(37) $(t)_p$ muß als Zeitpunkt interpretierbar sein.

m eine klarere Formalisierung zu erreichen, definiere ich:

(38) $t_i, t_j, t_k =_{Def}$ Zeiträume oder Zeitspannen
$t_m, t_n =_{Def}$ Zeitpunkte

enn wir diese Symbole nun in die Tiefenstrukturen einbringen, dann önnen wir folgendermaßen formuliern, wobei ich exemplarisch die iefenstruktur zu (35) angeben will:

(39) $(\exists x)[F(x).G(x).R(a,x).S(a,x,t_i)]^*(p),[T(b_1,t_n).T(b_2,t_n).T(b_3,t_n)]^*(q)$, wobei F = escalier, G = vermoulu, R = avoir, S = raccomoder, T = frapper à la porte, a = Grandet, b_{1-3} = les trois Cruchot

a hier im p-Ausdruck das Symbol für eine Zeitspanne t_i steht, ist ie Bedingung (37) nicht erfüllt, und (35) wird als ungrammatisch eingestuft. Für q könnte man auch schreiben:

$(\exists_{3!!}y)[H(y_1).H(y_2).H(y_3).T(y_1,t_n).T(y_2,t_n).T(y_3,t_n)]^{*}(q)$,
wobei H = membre de la famille Cruchot. $(\exists_{3!!}y)$ ist ein Numerator, zu lesen als: es existieren genau 3 y, für die gilt, daß ...
Die Schreibweise in (39) ist für unsere Zwecke jedoch ausreichend.

Auch (22), (23) und (36) verstoßen gegen die Bedingung (37).

Die Sätze (1) - (5), in denen 'alors que', 'pendant que' und 'tandis que' ebenso zu akzeptablen Ergebnissen führen wie 'dès que' und 'aussitôt que', lassen letztere eben deswegen zu, weil aufgrund des nicht vorhandenen Kontrasts zwischen imparfait und passé simple bzw. passé composé nicht eindeutig ausgeschlossen werden kann, daß es sich in $(S)_p$ auch um Zeitpunkte handeln kann. Statt beispielsweise der Tiefenstruktur (6) wäre also auch möglich:

(40) $[F(a,t_m)]^{*}(p),[G(b,t_n)]^{*}(q),(m=n)$,

womit die Bedingung (37) erfüllt ist. Ganz eindeutig ist (37) erfüllt in den Sätzen (24) - (26) für die wir zusammenfassend und ergänzend schreiben können:

(41) il fut parti, l'autre arriva.

(42) {Dès qu' / Aussitôt qu'} arrivé, il se coucha.

(43) vous voudrez, partez.

Dabei liegen unterschiedliche Typen von Zeitpunkten vor. In (41) und (42), das ja lediglich eine elliptische Konstruktion aus:

(44) {Dès qu' / Aussitôt qu'} il fut arrivé, il se coucha.

darstellt, handelt es sich um den Endpunkt einer Entwicklung in p, was man schematisch so darstellen könnte:

(45) ―――――――― . ,

während in (43) eher der Anfangspunkt nach dem Schema :

(46) . ――――――――

vorliegt. Entscheidend jedoch ist, daß ein Punkt zu fixieren ist. Dies ist allerdings genau der Grund, weshalb in Sätzen wie (41) - (43) die SUB-Elemente 'pendant que', 'alors que' und 'tandis que' nicht möglich sind. Innerhalb von p darf also hier kein fester Zeitpunkt fixiert sein, bzw. ein solcher Aspekt muß möglich sein. Die Bedingung wäre dann folgendermaßen zu formulieren:

(47) $(t)_p$ muß als Zeitspanne t_i interpretierbar sein.

Ob dabei in q ein Zeitpunkt oder eine Zeitspanne vorliegt, ist für die Möglichkeit des Einsatzes von 'alors que', 'pendant que' und 'tandis que' unerheblich:

(48) $\left\{\begin{array}{l}\text{Alors que}\\ \text{Pendant que}\\ \text{Tandis que}\end{array}\right\}$ Grandet raccomodait son escalier vermoulu, les trois Cruchot frappèrent à la porte.

(49) $\left\{\begin{array}{l}\text{Alors qu'}\\ \text{Pendant qu'}\\ \text{Tandis qu'}\end{array}\right\}$ il partait, nous arrivâmes.

(50) $\left\{\begin{array}{l}\text{Alors que}\\ \text{Pendant que}\\ \text{Tandis que}\end{array}\right\}$ sa famille s'épuise au travail, il fréquente les cafés.

ie genannten Funktionselemente können gesetzt werden, wenn entweder $t_i)_p = (t_j)_q$ oder wenn $(t_n)_q$ in $(t_i)_p$ enthalten sein kann. m das Gesagte zu verdeutlichen, gebe ich zu (48) und (50) die Tieenstrukturen an, indem ich die Temporalverhältnisse mitformuliere:

(39) $(\exists x)[F(x).G(x).R(a,x).S(a,x,t_i)]^{*}(p),[T(b_1,t_n).T(b_2,t_n).$ $.T(b_3,t_n)]^{*}(q),(n \text{ in } i)$

(8) $(\exists x)[F(x).R(a,x).G(x,t_i)]^{*}(p),(\exists y)[H(y).S(a,y,t_j)]^{*}(q),(i=j)$

an könnte natürlich Zeiträume auch als offene Klassen auffassen und eitpunkte als Elemente davon; oder man könnte Zeitspannen durch hre Endpunkte begrenzen und statt (n in i) schreiben: $(n \geq m)$ und $n \leq o)$, um auszudrücken, daß t_n in der Zeitspanne t_i, definiert urch den Anfangspunkt t_m und den Schlußpunkt t_o liegt. Jedoch ercheint es mir natürlicher, Zeiträume und Zeitpunkte als Temporalndividuen aufzufassen und durch entsprechende Indizierung zu unerscheiden, was nebenbei den Vorteil hat, daß dadurch die Notatinsweise erheblich vereinfacht wird.

Die Akzeptabilität von (3) und (16) liegt nun daran, daß zwei istinkte Tiefenstrukturen postuliert werden müssen:

(51) $(\exists x)[F(x).R(a,x).G(x,t_i)]^{*}(p),(\exists y)[H(y).S(a,y,t_j)]^{*}(q),(i=j)$

(52) $(\exists x)[F(x).R(a,x).G(x,t_m)]^{*}(p),(\exists y)[H(y).S(a,y,t_n)]^{*}(q),(m=n)$,

wobei F = famille, R = avoir, G = s'épuiser au travail, H = café, S = fréquenter, a = il, t_i, t_j = Zeitraumindividuen, t_m, t_n = Zeitpunktindividuen

Nachdem wir nun die Subjunktionsgruppen 'alors que', 'pendant que' nd 'tandis que' einerseits von 'aussitôt que' und 'dès que' anderrseits hinsichtlich ihres Gebrauchs abgegrenzt haben und die entprechenden Bedingungen formuliert haben, ist festzustellen, daß es och Unterschiede innerhalb der Gruppen gibt. So sind 'aussitôt que' nd 'dès que' nicht funktionsgleich, obwohl sie in den meisten Gramatiken praktisch als Synonyme behandelt werden. So ist beispielsweise folgender Satz mit 'dès que' akzeptabel, während er mit 'aussitôt que' abgelehnt wird:

(53) Dès qu'il est parti, nous sommes sans nouvelles.

(54) *Aussitôt qu'il est parti, nous sommes sans nouvelles.

Ich habe mich mit dem Problem noch nicht eingehender befaßt, jedoch scheint mir (54) daran zu scheitern, daß $(t)_q$ eindeutig ein Zeitraum ist, d. h. als ein Zeitpunkt nicht interpretierbar. Dazu die Tiefenstruktur:

(55) $[F(a,t_m)]^*(p), (\exists_2 x)\overline{(\exists y)}[G(y).(R(x_1,y,t_j) v R(x_2,y,t_j) v \ldots v R(x_k,y,t_j)).\overline{I(x_1,x_2)}]^*(q)$, wobei F = ⌜être parti⌝, G = ⌜nouvelle⌝, R = avoir, I = identisch, a = il, x_1 = je, x_2, x_k = unbestimmte Individuen

Die Bedingung für den Einsatz von 'aussitôt que' könnte demnach ergänzend zu (37) sein:

(56) $(t)_q$ muß als Zeitpunkt t_n interpretierbar sein.

Es ist allerdings auch eine andere Erklärung denkbar, nämlich die, daß 'dès que' nicht unbedingt die Gleichzeitigkeit benötigt, sondern wie 'depuis que' die Vorzeitigkeit zuläßt, was für 'aussitôt que' nicht gilt. Da ich noch kein hinreichend großes überprüftes Corpus dazu habe, möchte ich dieser Frage hier nicht weiter nachgehen, sondern auf eine andere Schwierigkeit hinweisen.

REINHART MÜLHAUSE schreibt in seinem Buch "Grundprobleme der französischen Grammatik" im Zusammenhang mit 'pendant que' und 'tandis que', daß ersteres "zeitlich", letzteres "gegenüberstellend" sei. Nach einigen Beispielsätzen schreibt er dann: "In Zweifelfällen verwende man 'alors que', das sowohl zeitliche als auch gegenüberstellende Bedeutung haben kann."

Ohne daß ich auf die Formulierung selbst eingehen möchte, muß ich aber feststellen, daß die Behauptung, die 'alors que' gewissermaßen als Patentlösung vorschlägt, nicht zutrifft, denn sie wird folgenden Beispielen nicht gerecht:

(57) {Pendant qu' / Tandis qu'} il y a de la vie, il y a de l'espoir.

(58) *Alors qu'il y a de la vie, il y a de l'espoir.

(59) {Pendant que / Tandis que} nous serons dans la voiture, je mettrai les souliers de réchange.

(60) *Alors que nous serons dans la voiture, je mettrai les souliers de réchange.

(61) Alors qu'elle avait payé, elle lui dirait tout.

(62) *$\left\{\begin{array}{l}\text{Pendant qu'}\\ \text{Tandis qu'}\end{array}\right\}$ elle avait payé, elle lui dirait tout.

(63) Alors qu'il m'était arrivé de prendre une direction, je tâchais de la suivre.

(64) *$\left\{\begin{array}{l}\text{Pendant qu'}\\ \text{Tandis qu'}\end{array}\right\}$ il m'était arrivé de prendre une direction, je tâchais de la suivre.

(65) $\left\{\begin{array}{l}\text{Alors qu'}\\ \text{Tandis qu'}\end{array}\right\}$ il est parti, nous sommes sans nouvelles.

(66) *Pendant qu'il est parti, nous sommes sans nouvelles.

(67) $\left\{\begin{array}{l}\text{Alors qu'}\\ \text{Tandis qu'}\end{array}\right\}$ elle m'a vu, ces mots ont éclaté d'un transport imprévu.

(68) *Pendant qu'elle m'a vu, ces mots ont éclaté d'un transport imprévu.

Wäre MÜLHAUSEs Behauptung richtig, müßte 'alors que' immer zu akzeptablen Ergebnissen führen, indem es einmal mit 'pendant que', ein anderes Mal mit 'tandis que' zusammenginge. Die Sätze (57) - (68) lassen sich unschwer in drei Gruppen einteilen. In der ersten (57) - (60) wird 'alors que' nicht akzeptiert; in der zweiten (61) - (64) 'pendant que' und 'tandis que' und in der dritten (65) - (68) wird 'pendant que' nicht akzeptiert, während hier 'alors que' und 'tandis que' zusammengehen. Fangen wir mit der letzten Gruppe an, auf die MÜLHAUSEs Behauptung noch am ehesten zutrifft. Die Sätze (66) und (68) werden wohl deshalb nicht akzeptiert, weil die Bedingung

(47) $(t)_p$ muß als eine Zeitspanne t_i interpretierbar sein.

nicht erfüllt ist, da es sich doch eher um Zeitpunkte handelt, was durch die Tempuswahl gekennzeichnet ist. Außerdem wäre besonders in (66) eine Gleichzeitigkeit logisch nicht konstruierbar, selbst wenn $(t)_p$ eine Zeitspanne wäre. Dagegen werden (69) und (70) akzeptiert:

(69) Pendant qu'il est absent, nous sommes sans nouvelles.

(70) Pendant qu'elle me voyait, ces mots ont éclaté d'un transport imprévu.

mit den Zeitverhältnissen $(t_i)_p = (t_j)_q$ in (69) und $(t_n)_q$ in $(t_i)_p$ in (70). Nun hatten wir (47) auch für 'alors que' und 'tandis que' formuliert. Wir können dies in dieser Form nicht aufrechterhalten, weil sonst (65) und (67) ausgeschlossen würden. Tatsächlich können 'alors que' und 'tandis que' unter zwei verschiedenen Gesichtspunkten eingeführt werden, dem der Gleichzeitigkeit wie 'pendant que' nach den Maßgaben von (47) und dem der Gegensätzlichkeit, wobei ein Gesichtspunkt genügt. Folgerichtigerweise ist ein Satz wie:

(71) *Tandis qu'il arrive à la maison, il part tout de suite.

auszuschließen, da weder Gleichzeitigkeit noch Gegensätzlichkeit gegeben sind. Dabei scheint 'alors que' sogar den Gegensatz bzw. seine Interpretierbarkeit zu fordern, was bei 'tandis que' offenbar nicht der Fall ist, wie sich an (58) und (60) manifestiert, wo zwar Gleichzeitigkeit gegeben ist, ein Gegensatz aber wenigstens in normalem Kontext kaum interpretierbar erscheint. Für 'tandis que' genügt die Gleichzeitigkeit, wie aus der Akzeptabilität von (57) und (59) hervorgeht. Dies findet übrigens eine Stütze in der Aussage des PETIT ROBERT, daß 'alors que' inzwischen in seinem temporalen Gebrauch veraltet sei, während der gegensätzliche Gebrauch durchaus lebendig sei.

Ein Problem stellen nun noch die Sätze (61) und (63) insofern dar als hier 'alors que' akzeptiert wird. Der Umstand, daß (62) und (64) nicht akzeptiert werden, ist regelkonform, da weder (47) erfüllt ist (und selbst wenn dies der Fall wäre, dann würde die Gleichzeitigkeit auf der Strecke bleiben), noch ein Gegensatz vorliegt. Im Gegenteil könnte man eher, insbesondere in (63) von einer Grund-Folge-Beziehung ausgehen. Ich nehme an, ohne daß mich der Lösungsweg bisher voll befriedigt, daß es sich in (61) und (63) nicht um dasselbe 'alors que' handelt wie in den übrigen Sätzen, sondern eher um ein 'alors', das eigentlich zu $(S)_q$ gehört und dem ein durch 'que' eingeleiteter Satz folgt, etwa in dem Sinne von 'maintenant, que ... ': 'nun, da ... ' oder 'jetzt, wo ... '. Eine solche Lesart ist für 'pendant que' und 'tandis que' ausgeschlossen. Eine eingehendere Untersuchung dazu ist freilich notwendig.

Die Regeln, nach denen die Sätze gebildet werden, haben folgendes Aussehen:

(T 1) $SUB_1 + (S)_{2\,p} + (S)_{3\,q} \rightarrow$ tandis que_1 + 2 + 3

Bed.: $(((t)_p$ als Zeitraum t_i interpretierbar$).((i=j)v(n$ in $i)))v^p\mathfrak{M}(p,q)$

Dabei verstehe ich $\mathfrak{M}(p,q)$ als eine abkürzende Schreibweise von p mais q. Das bedeutet, daß der Funktor '$\mathfrak{M}$' einen Spezialfall der Und-Verknüpfung symbolisiert, der zwischen den Aussagen p und q operiert und der asymmetrisch, nicht transitiv, irreflexiv und nur dann wahr ist, wenn p.q wahr ist.

(T 2) $SUB_1 + (S)_{2\,p} + (S)_{3\,q} \rightarrow$ dès que_1 + 2 + 3

Bed.: $(t)_p$ als Zeitpunkt t_m interpretierbar

us (T 1) bzw. (T 2) leite ich die übrigen SUB-Elemente auf trans-
ormationellem Wege her:

(T 3) $\underset{1}{\text{tandis que}} + \underset{2}{(S)_p} + \underset{3}{(S)_q} \;\; \text{op}\!\rightarrow\;\; \underset{1}{\text{alors que}} + 2 + 3$

Bed.: $\mathfrak{M}(p,q)$ interpretierbar

(T 4) $\underset{1}{\text{tandis que}} + \underset{2}{(S)_p} + \underset{3}{(S)_q} \;\; \text{op}\!\rightarrow\;\; \underset{1}{\text{pendant que}} + 2 + 3$

Bed.: $((t)_p$ als Zeitraum t_i interpretierbar$).((i=j) v (n$ in $i))$

(T 5) $\underset{1}{\text{dès que}} + \underset{2}{(S)_p} + \underset{3}{(S)_q} \;\; \text{op}\!\rightarrow\;\; \underset{1}{\text{aussitôt que}} + 2 + 3$

Bed.: $(t)_p$ und $(t)_q$ als Zeitpunkte t_m und t_n interpretierbar, oder aber es ist die Bedingung in (T 2) zu ändern in: $((t)_p$ als Zeitpunkt t_m interpretierbar$).((t_m \leq t_n) v (t_m \leq t_j))$, was wie gesagt noch genauer zu untersuchen wäre.

ine fakultative Transformation permutiert die Reihenfolge der Sätze:

(T 6) $$\underset{1}{\left\{\begin{array}{l}\text{tandis que}\\ \text{alors que}\\ \text{pendant que}\\ \text{dès que}\\ \text{aussitôt que}\end{array}\right\}} + \underset{2}{(S)_p} + \underset{3}{(S)_q} \;\; \text{op}\!\rightarrow\;\; 3 + 1 + 2$$

amit werden Sätze erzeugt wie:

(72) On fait des discours, pendant qu'il faut agir.

(73) Il fréquente les cafés, tandis que sa famille s'épuise au travail.

aneben ist noch der Einschub möglich nach dem Muster:

(74) Les trois Cruchot, dès que Grandet raccomodait son escalier vermoulu, frappèrent à la porte.

ie Regel dazu hat folgendes Aussehen:

(T 7) $$\underset{1}{\left\{\begin{array}{l}\text{tandis que}\\ \text{alors que}\\ \text{pendant que}\\ \text{dès que}\\ \text{aussitôt que}\end{array}\right\}} + \underset{2}{(S)_p} + (\underset{3}{NP} + \underset{4}{(ne)} + \underset{5}{V} + \underset{6}{AUX} + \underset{7}{(pas)} + \underset{8}{X})_q$$

$\text{op}\!\rightarrow\;\; 3 + 1 + 2 + 4 + 5 + 6 + 7 + 8$

Bed.: 3[-PRO]

ie Bedingung in (T 7) schließt ungrammatische Bildungen aus wie:

(75) *Il, tandis que sa famille s'épuise au travail, fréquente les cafés.

T 7) kommt mit steigender Länge von $(S)_p$ immer seltener zur Anwen-
ung.

GIBT ES KONTRAFAKTIVE VERBEN?

Maria-Elisabeth Conte

> Hypotheses non fingo.
> Isaac Newton

> Where much is alleged
> something must be true.
> Edward Gibbon

1.1. In ihrem bekannten Aufsatz "Fact" haben Carol und Paul Kiparsky Prädikate mit Satzkomplementen untersucht und dabei die faktiven von den nicht-faktiven Prädikaten unterschieden. Faktive Prädikate sind dadurch gekennzeichnet, dass der Sprecher die im Komplementsatz ausgedrückte Proposition als wahr präsupponiert. Es handelt sich um eine Forschung, in der der Sinn der faktiven Prädikate nicht an sich, sondern im Verhältnis zur Bedeutung im Sinne Gottlob Freges (d.h. zum Wahrheitswert) des Komplementsatzes untersucht wird.

1.2. Lauri Karttunen hat die von den Kiparskys bloss _ex negativo_ definierte Klasse der nicht-faktiven Prädikate positiv spezifiziert, indem er Subklassifizierungen für sie erarbeitet hat. Dabei beschränkt er sich weitgehend auf Verben (KARTTUNEN 1970, 1971a, 1971b). So wie bei C. und P. Kiparsky beruht auch bei Karttunen die Typologie der nicht-faktiven Verben mit Satzkomplementen auf der Beziehung zwischen Verben und dem Wahrheitswert des Komplementsatzes.

1.2.1. Karttunen führt hierzu die Beziehung der Implikation ein: im Mittelpunkt seiner Typologie stehen eben die von ihm so bezeichneten "implicative verbs". Sie drücken eine notwendige und zureichende Bedingung für die Wahrheit des Komplementsatzes aus (weshalb Karttunen sie auch _if-and-only-if-verbs_ nennt). Im Unterschied zu den faktiven Verben, bei denen der Komplementsatz eben das _présupposé_ ist, das von der Negation nicht angetastet wird, gehört der Komplementsatz bei den implikativen Verben zum _posé_, und fällt somit unter den Skopus der Negation. (Die Termini _posé_ und _présupposé_ übernehme ich von DUCROT 1972.)

Implikative Verben sind z.B. <u>to manage</u>, <u>to dare</u>, <u>to succeed</u>. Der Satz (1)

(1) Es gelang ihm, den Zug vor der Brücke zu stoppen.

impliziert den Satz (2)

(2) Er stoppte den Zug vor der Brücke.

Die Verneinung von (1), der Satz (3)

(3) Es gelang ihm nicht, den Zug vor der Brücke zu stoppen.

impliziert dagegen den Satz (4)

(4) Er stoppte den Zug nicht vor der Brücke.

Karttunen macht darauf aufmerksam, dass es sich hier nicht um Implikation im logischen Sinne handelt. Vielmehr gebraucht er <u>to imply</u> in dem Sinne, dass der Satz <u>p</u> den Satz <u>q</u> impliziert, wenn die Behauptung von <u>p</u> den Sprecher auf <u>q</u> festlegt. Bei Karttunen heisst es: "'<u>p</u> implies <u>q</u>' means only that asserting <u>p</u> commits the speaker to <u>q</u>" (KARTTUNEN 1971a: 344).[1] Der Sprecher ist auf die Wahrheit bzw. auf die Falschheit der durch den Komplementsatz ausgedrückten Proposition für die Folge seiner Rede festgelegt. Es würde z.B. ein Widerspruch sein, zu sagen:

(5) +Es gelang ihm, den Zug vor der Brücke zu stoppen.
Aber der Zug hielt nicht.

1.2.2. Von den so definierten implikativen Verben, die zweiseitig implikativ sind, unterscheidet Karttunen diejenigen Verben, die nur einseitig implikativ sind: die <u>if-verbs</u> (Wenn-Verben) und die <u>only-if-verbs</u> (Nur-wenn-Verben).

Wenn-Verben, die eine zureichende Bedingung ausdrücken, sind z.B. <u>to cause</u> und <u>to force</u>. Nur-wenn-Verben, die eine notwendige Bedingung ausdrücken, sind z.B. <u>to be able</u>, <u>to be in the position</u>. Beim positiven Gebrauch eines Satzes mit einem Wenn-Verb ist der Sprecher auf die Wahrheit des Komplementsatzes festgelegt. Beim negativen Gebrauch ist der Sprecher auf keinen Wahrheitswert festgelegt: die Implikation ist neutral.

Während der Satz (6)

(6) Er zwang die Reisenden, im Abteil zu bleiben.

den Satz (7)

(7) Die Reisenden blieben im Abteil.

impliziert, impliziert der Satz (8)

(8) Er zwang die Reisenden nicht, im Abteil zu bleiben.

weder die Wahrheit noch die Falschheit von (7).

Die Nur-wenn-Verben verhalten sich genau umgekehrt.[2]

1.2.3. Diesen drei positiven Verbklassen stellt Karttunen nun jeweils eine negative Verbklasse zur Seite: neben den implikativen Verben gibt es die negativ implikativen Verben (z.B. <u>to fail</u>), neben den Wenn-Verben gibt es die negativen Wenn-Verben (z.B. <u>to prevent</u>), neben den Nur-wenn-Verben gibt es die negativen Nur-wenn-Verben (z.B. <u>to hesitate</u>).

Ich führe ein Satzpaar mit einem negativen Wenn-Verb an, denn auf eben diese Verbklasse werde ich später noch zurückkommen.

(9) Er verhinderte, dass das Unglück geschah.

impliziert

(10) Das Unglück geschah nicht.

Die Verneinung von (9), der Satz (11)

(11) Er verhinderte nicht, dass das Unglück geschah.

impliziert dagegen weder die Wahrheit noch die Falschheit von (10). Bei (9) ist der Sprecher auf die Falschheit des Komplementsatzes festgelegt. Den Satz (11) kann er dagegen ohne Selbstwiderspruch auf zweifache Weise fortsetzen:

(12) Er verhinderte nicht, dass das Unglück geschah. Er verhinderte nur, dass zu viele Menschen dabei umkamen, indem er einige Wagen des Zuges vor der Brücke abhängte.

(13) Er verhinderte nicht, dass das Unglück geschah. Der Zug hielt von selbst vor der gefährlichen Brücke.

Sowohl wenn der Sprecher in der Fortsetzung seiner Rede behauptet, dass das Unglück geschah (wie in (12)), als auch wenn er behauptet, dass das Unglück nicht geschah (wie in (13)), entsteht ein widerspruchsfreier Text.

1.3. Rein kombinatorisch hat nun Karttunen auch den faktiven Verben ein negatives Pendant zugeordnet: die <u>counterfactive verbs</u>. Die kontrafaktiven Verben wurden in KARTTUNEN 1970 <u>en passant</u> eingeführt, und das einzige Beispiel, das Karttunen für die Klasse der kontrafaktiven Verben anführt, ist <u>to pretend</u> (<u>vorgeben</u>, <u>vortäuschen</u>, <u>so tun als ob</u>). Die Hypothese der Kontrafaktivität kann man folgenderweise formulieren: es gibt Verben, die dadurch gekennzeichnet sind, dass der Sprecher die im Komplementsatz ausgedrückte Proposition als falsch präsupponiert.

Der Sprecher präsupponiert z.B. bei der Äusserung des Satzes:

(14) Der Student täuschte vor, die Klausur zu schreiben.

die Falschheit von (15)

(15) Der Student schrieb die Klausur.

Die Hypothese der Kontrafaktivität sieht nicht unbegründet aus: offensichtlich besteht ein Verhältnis zwischen (14) und der Falschheit von (15).

1.3.1. Und in der Tat ist die Hypothese der Kontrafaktivität in der linguistischen Literatur mehrfach rezipiert worden.

1.3.1.1. Verschiedene Autoren übernehmen die Kategorie der kontrafaktiven Verben: Luise F. Pusch, Herbert E. Brekle, Eva Hajičová, Steven E. Boër und William G. Lycan. Die Bezeichnung schwankt zwischen 'kontrafaktiven Verben' und 'negativ-faktiven Verben'.[3]

1.3.1.2. Auffällig ist, dass Ole Mørdrup in einer umfangreichen Arbeit über die Semantik der französischen Verben mit Satzkomplementen (MØRDRUP, 1975) unabhängig von Karttunen negativ-faktive Verben einführt. Zwar bezieht sich Mørdrup weitgehend auf Karttunen, scheint aber zufällig nicht den Aufsatz: On the Semantics of English Complement Sentences (KARTTUNEN 1970) zu kennen, in dem Karttunen eben die counterfactive verbs eingeführt hat. Nach Mørdrup vervollständigen die negativ-faktiven Verben das logische Schema von Karttunen: "Karttunen n'a pas mentionné cette classe dans sa description des verbes anglais. Ils complètent pourtant les possibilités logiques de son schéma." (MØRDRUP 1975: 137).

1.3.2. Welche Verben gehören nun zur Klasse der angeblich kontrafaktiven Verben? Bei Karttunen finden wir, wie schon bemerkt, nur to pretend. Luise F. Pusch gibt to pretend mit vorgeben wieder.

1.3.2.1. Herbert Ernst Brekle fügt ein anderes Verb hinzu: neben vortäuschen gibt er lügen an. Aber die Hinzufügung von lügen ist falsch.

Gegen die Einstufung von lügen als kontrafaktives Verb gibt es zwei Argumente: ein syntaktisches und ein semantisches.

1.3.2.1.1. Syntaktisches Argument: Die Kontrafaktivität von lügen ist a priori durch die Valenz (Wertigkeit) von lügen ausgeschlossen: lügen hat keinen Komplementsatz und daher kann auch die Falschheit des Komplementsatzes nicht präsupponiert werden.

Ein Satz wie (16)

(16) +S hat gelogen, dass p.

wäre syntaktisch nicht wohlgeformt, wäre dystaktisch. Die Kontrafaktivität von lügen ist also schon wegen der Valenz des Verbs ausgeschlossen.

1.3.2.1.2. Semantisches Argument: Die Falschheit dessen, was bei der Lüge gesagt wird, ist nicht das présupposé. Z.B. im Satz (17)

(17) Epimenides lügt.

ist die Falschheit dessen, was Epimenides sagt, das posé und nicht das présupposé.

Fazit der beiden Argumente: Lügen ist kein kontrafaktives Verb.

1.3.2.2. Bei Mørdrup finden wir eine längere Liste von angeblich kontrafaktiven Verben (die auch von KÖRNER 1977 zitiert wird): affecter de, feindre de, simuler de, se figurer, s'imaginer, rêver.[4]

1.3.2.3. Zur Gruppe der Verben, die sich so wie die angeblich kontrafaktiven Verben verhalten, gehören im Deutschen ausser vorgeben, vortäuschen, so tun als ob noch: simulieren, sich einbilden, wähnen, in der Illusion leben, jemandem etwas weismachen, jemandem etwas vormachen.[5]

1.3.3. Nach der Hypothese der Kontrafaktivität ist die Beziehung zwischen einem Satz mit einem kontrafaktiven Verb und seinem Komplementsatz eine Präsuppositionsbeziehung. Aber: Handelt es sich wirklich um Präsupposition? Und: Sind die angeblich kontrafaktiven Verben wirklich das negative Pendant zu den faktiven Verben?

2.1. Präsuppositionen bleiben bekanntlich konstant (unangetastet) auch wenn ein Satz negiert wird. Bei Sätzen mit einem faktiven Verb besteht sowohl beim positiven als auch beim negativen Satz die Präsupposition der Wahrheit des Komplementsatzes.

2.1.1. Besteht nun symmetrisch die Präsupposition der Falschheit des Komplementsatzes (die negative Präsupposition oder Kontrapräsupposition[6]) sowohl bei einem positiven als auch bei einem negativen Satz mit einem angeblich kontrafaktiven Verb? Gelingt der Negationstest bei den angeblich kontrafaktiven Verben? Wird sowohl bei (14) als auch bei (18) die Falschheit von (15) präsupponiert?

Vergleichen wir die drei Sätze:

(14) Der Student täuschte vor,die Klausur zu schreiben.

(18) Der Student täuschte nicht vor, die Klausur zu schreiben.

(15) Der Student schrieb die Klausur.

Bei dem negativen Satz (18) bleibt die angebliche Präsupposition der Falschheit (die Kontrapräsupposition) nicht erhalten: der Negationstest gelingt nicht.[7]

2.1.2. Der negative Satz (18) hat je nach Kontext und Betonung zwei Lesarten.

2.1.2.1. In der ersten (privilegierten) Lesart bedeutet (18), dass der Student die Klausur tatsächlich schrieb. (Mit anderen Worten: er schrieb die Klausur nicht zum Schein, sondern in der Wirklichkeit.) In dieser ersten Lesart ist das <u>posé</u> von (18) genau die Negation seiner angeblichen Kontrapräsupposition (d.h. das <u>posé</u> von (18) ist die Negation der Falschheit von (15)). Die angebliche Kontrapräsupposition besteht nicht.

2.1.2.2. In der zweiten Lesart bedeutet (18), dass es nicht der Fall ist, dass der Student <u>vortäuschte</u>, die Klausur zu schreiben. Hier wird der Nichtvollzug der Handlung des Vortäuschens behauptet. Auch bei dieser zweiten Lesart von (18) besteht nicht die angebliche Kontrapräsupposition.

2.1.2.3. Fazit: In keiner der beiden Lesarten besteht die angebliche Kontrapräsupposition; die Falschheit des Komplementsatzes wird nicht präsupponiert.[8]

2.2. Ein möglicher Einwand dagegen, dass das Nicht-Gelingen des Negationstests das Vorhandensein einer Präsuppositionsbeziehung ausschliesst, könnte nun lauten: Auch wenigstens in einem anderen Fall, in dem man von Präsupposition spricht, gelingt der Negationstest nicht; aber dennoch handelt es sich um Präsupposition.

2.2.1. Es ist der Fall der Ergänzungsfragen. Bei einer Ergänzungsfrage wie:

(19) Was hat Peter getan?

präsupponiert der Sprecher, dass Peter etwas getan hat. Hierzu bemerkt nun John Lyons, dass "this kind of presupposition does not remain constant under negation" (LYONS 1977: 597). Für die Frage:

(20) Was hat Peter nicht getan?

besteht nämlich nicht die Präsupposition, dass Peter etwas getan hat, sondern vielmehr die Präsupposition, dass es etwas gibt, was Peter nicht getan hat.

2.2.2. Aber dieser hypothetische Einwand ist unbegründet. Dass im Falle des Satzpaares (19) und (20) bei der Negation die Präsupposition nicht konstant bleibt, lässt sich meiner Meinung nach anders erklären: (20) ist nicht die Verneinung von (19), sondern lediglich die negative Frage zu (19). Und es könnte nicht anders sein: denn (a) keine Frage ist die Negation einer Frage; (b) Fragen können zwar negativ sein, aber können nicht verneint werden.[9] Die negative Frage ist n i c h t die Verneinung der positiven Frage, sondern einfach eine andere Frage als die positive (eine eigenständige Frage mit ihren eigenen Präsuppositionen).[10] Der hypothetische Einwand ist also unbegründet, da bei den Ergänzungsfragen der Negationstest von vornherein ausgeschlossen ist.

2.2.3. Im Unterschied zu den Ergänzungsfragen ist ein Satz mit einem angeblich kontrafaktiven Verb (wie (14)) negierbar (die Negation von (14) ist (18)). Also ist der Negationstest hier anwendbar, um zu entscheiden, ob eine Präsupposition, bzw. eine Kontrapräsupposition, besteht.

2.3. Wenn aber bei den angeblich kontrafaktiven Verben die Präsupposition der Falschheit (die Kontrapräsupposition) nicht besteht, dann handelt es sich bei diesen Verben nicht um das negative Pendant zu den faktiven Verben: dann sind die angeblich kontrafaktiven Verben nicht kontrafaktiv.

Wenn aber die angeblich kontrafaktiven Verben nicht kontrafaktiv sind, was für Verben sind es dann? Und wenn es sich nicht um die Relation der Präsupposition (insbesondere der Kontrapräsupposition) handelt, was für eine Relation besteht dann zwischen den angeblich kontrafaktiven Verben und dem Wahrheitswert des Komplementsatzes?

3.1. Was für Verben sind nun die als kontrafaktiv klassifizierten Verben?

3.1.1. Betrachten wir nochmal die beiden Sätze (14) und (18). Wozu verpflichten sie den Sprecher, der sie äussert? Worauf legen sie ihn fest? Und welche folgenden Äusserungen sind mit der

Äusserung von (14) und (18) jeweils kompatibel? Bei der Äusserung von (14) ist der Sprecher dazu committed, es für falsch zu halten, dass der Student die Klausur geschrieben hat. (Wenn ein Sprecher äussern würde:

(21) +Der Student täuschte vor, die Klausur zu schreiben. Aber er schrieb die Klausur.

dann wäre seine Rede unkohärent. Ein solcher Text wäre offensichtlich selbstwidersprüchlich.)

Durch die Äusserung von (18) besteht für den Sprecher dagegen überhaupt kein commitment in bezug auf die Wahrheit des Komplementsatzes. Eben weil der Satz zwei Lesarten zulässt, ist der Sprecher weder darauf festgelegt, es für falsch zu halten, noch es für wahr zu halten, dass der Student die Klausur geschrieben hat. In der Tat sind mit dem Satz (18) zwei miteinander unverträgliche Fortsetzungen verträglich. Z.B. sind sowohl (22) als auch (23) widerspruchsfreie Texte:

(22) Der Student täuschte nicht vor, die Klausur zu schreiben. Ganz im Gegenteil, er schrieb die Klausur höchst konzentriert und löste sehr schnell alle Aufgaben.

(23) Der Student täuschte nicht vor, die Klausur zu schreiben. Er verzichtete gleich zu Anfang darauf und löste in der Zeit ein Kreuzworträtsel.

3.1.2. Bei einem Vergleich des Verhaltens der angeblich kontrafaktiven Verben mit den zu Anfang dieser Arbeit eingeführten Klassifizierungen von Karttunen ergibt sich, dass vortäuschen sich genauso verhält wie die negativen Wenn-Verben.

Bei ihrem positiven Gebrauch legt der Sprecher sich darauf fest, dass die im Komplementsatz ausgedrückte Proposition falsch ist. Bei ihrem negativen Gebrauch besteht kein commitment für den Sprecher in bezug auf die Wahrheit des Komplementsatzes. Genauso wie der Satz (9)

(9) Er verhinderte, dass das Unglück geschah.

den Satz (10) impliziert

(10) Das Unglück geschah nicht.

so impliziert der Satz (14)

(14) Der Student täuschte vor, die Klausur zu schreiben.

den Satz (24)

(24) Der Student schrieb die Klausur nicht.

Und genauso wie die Verneinung von (9), d.h. der Satz (11)

(11) Er verhinderte nicht, dass das Unglück geschah.
weder die Wahrheit noch die Falschheit von (10) impliziert, genauso impliziert der Satz (18)

(18) Der Student täuschte nicht vor, die Klausur zu schreiben.
weder die Wahrheit noch die Falschheit von (24).

3.1.3. Auf die am Ende von 2. gestellte Frage: Was für Verben sind nun die fälschlicherweise als kontrafaktiv klassifizierten Verben? kann ich also antworten: Sie gehören zur Klasse der negativen Wenn-Verben. So unerwartet es klingen mag: Vom Standpunkt dieser Verbklassifikation aus (nämlich vom Standpunkt der Beziehung der Verben zum Wahrheitswert ihres Komplementsatzes) rückt to pretend in die gleiche Verbklasse wie to prevent (und das nicht wegen der Lautähnlichkeit). Beide Verben sind negative if-verbs.

3.2. Und nun zur zweiten der am Ende von 2. formulierten Fragen: Was für eine Relation besteht zwischen einem Verb wie vortäuschen (das ich als negatives if-verb klassifiziert habe) und dem Wahrheitswert des Komplementsatzes?

3.2.1. Es ist nicht die Beziehung der Präsupposition (insbesondere nicht die Kontrapräsuppositionsbeziehung), sondern vielmehr die Beziehung der allegation. Dieser Terminus wurde von Eva Hajičová eingeführt, die allegation von meaning und presupposition unterscheidet. Es handelt sich um allegation "when under negation there is only a potential possibility of what is true in the affirmative statement" (SGALL/HAJIČOVÁ/BENEŠOVÁ 1974: 104).

3.2.2. Die Definition der allegation trifft genau auf die Wenn-Verben und auf die negativen Wenn-Verben zu. Und in der Tat enthält Hajičovás Liste der Verben die durch allegation gekennzeichnet sind, Wenn-Verben (to cause, to force) und negative Wenn-Verben (to prevent, to dissuade, to keep from, to discourage).

3.2.3. Dieser Menge von Verben, die durch allegation gekennzeichnet sind, füge ich nun die Menge der angeblich kontrafaktiven Verben hinzu, die (vom Standpunkt ihres Verhältnisses zum Wahrheitswert des Komplementsatzes) wie die negativen Wenn-Verben sind und schlage vor, die Verben, bei denen die Beziehung der allegation besteht, allegative Verben zu nennen.

3.3. Auf die Frage des Titels meiner Arbeit: _Gibt es kontrafaktive Verben?_ gebe ich folgende Antwort: Meiner These nach sind alle angeblich kontrafaktiven Verben allegative Verben.
Die Frage aber, ob kontrafaktive Verben überhaupt unmöglich sind, ist eine andere Frage. Wenn ich auf diese zweite Frage antworten würde, kontrafaktive Verben seien überhaupt unmöglich, dann wäre meine Antwort eine unerwiesene Behauptung, eine _allegation_.

ANMERKUNGEN

1 Das _commitment_, das im Definiens von Karttunens _implication_ vorkommt, ist nicht das deontische _commitment_, das bekannterweise in der deontischen Logik untersucht worden ist (für die Literatur vgl. CONTE/DI BERNARDO 1977), sondern ein doxastisches _commitment_. Karttunen schreibt: "_p_ implies _q_ iff whenever _p_ is asserted the speaker ought to believe _q_" (KARTTUNEN 1971b: 6).

2 Bei den Nur-wenn-Verben ist die Implikation beim positiven Gebrauch neutral, während der Sprecher beim Gebrauch eines negativen Satzes mit einem Nur-wenn-Verb auf die Falschheit des Komplementsatzes festgelegt ist.

3 Bei PUSCH 1972 heissen sie 'kontrafaktive Verben'; bei BREKLE 1972 'negativ-faktive Verben'; bei HAJIČOVÁ 1974 'negative factive verbs'; bei MØRDRUP 1975 'verbes factifs négatifs'; bei BOËR/LYCAN 1976 'counterfactive verbs'; bei KÖRNER 1977 'verbes factifs négatifs'.

4 _Rêver_ (_träumen_) ist ein weltkonstitutives Verb. Ob auch die anderen angeblich kontrafaktiven Verben weltkonstitutiv sind, ist eine Frage, der ich hier nicht nachgehe.

5 Auch einige Verben ohne Komplementsatz, die kristallisierte idiomatische Ausdrücke sind, verhalten sich so wie die angeblich kontrafaktiven Verben. Z.B.: _sich ausgeben als_ (z.B.: Er gab sich als Arzt aus), _spielen_ (z.B.: Er spielt den Dummen), _sich stellen_ (z.B.: Er stellte sich unwissend).
Noch ein Kommentar zu dem Verb _vorgeben_: In einer ganz neuen Arbeit von János S. Petőfi und Hermann Kayser über performative und weltkonstitutive Ausdrücke des Deutschen wird das Verb _vorgeben_ zu den performativen Verben gerechnet. Mir scheint aber, dass für ein Verb wie _vorgeben_ (wie übrigens für die anderen angeblich kontrafaktiven Verben) die performative Verwendung ausgeschlossen ist. Es ist unmöglich, dadurch etwas vorzugeben, dass man sagt: "Ich gebe vor, dass ...". Die Handlung des Vorgebens würde nämlich von vornherein dadurch

vereitelt, dass der Sprecher selbst seine Handlung als ein Vorgeben entlarvt. (Analogerweise kann man sich nicht dadurch das Recht, als erster zu sprechen, anmassen, dass man sagt: "Hiermit masse ich mir das Recht an, als erster zu sprechen". Es wäre ein pragmatisches Unding, eine pure Dyspraxie.) Es muss allerdings bemerkt werden, dass Petőfi und Kayser einen anderen Performativitätsbegriff verwenden als den in der Sprachphilosophie und Linguistik üblichen (vgl. PETŐFI/ KAYSER 1977).

6 Das Wort 'Kontrapräsupposition' ist mir vielleicht durch das Wort 'contrapositio' ('Kontraposition') suggeriert worden, das von dem Philosophen geprägt wurde, der eben am Ort unseres 12. Linguistischen Kolloquiums, in Pavia, starb und der auch in Pavia begraben liegt: Severinus Boethius (vgl. PRANTL 1855: 698).

7 Entsprechend ist das Ergebnis des Fragetests.

8 Auch Mørdrup bemerkt, dass der negative Satz zwei Lesarten zulässt. Er unterscheidet interne und externe Negation. Bei der internen Negation wird seiner Meinung nach die Falschheit des Komplementsatzes präsupponiert; bei der externen Negation kann ein Satz mit einem negativ-faktiven Verb "impliquer que le complément est vrai" (MØRDRUP 1975: 138).
Bei der einen Lesart von Präsupposition der Falschheit und bei der anderen Lesart von Implikation der Wahrheit zu sprechen und dennoch die Kategorie der negativ-faktiven Verben beizubehalten, die ja die Eigenschaft haben sollen "de présupposer que le complément est faux" (MØRDRUP 1975: 137), scheint mir unfolgerichtig.

9 Hier taucht das Problem auf, ob die ἀπόφασις nur im Falle der ἀπόφανσις möglich ist, d.h. ob die Verneinung nur beim λόγος ἀποφαντικός möglich ist.

10 Wenn überhaupt könnte man metaphorisch die Zurückweisung der Präsupposition einer Frage als Verneinung der Frage betrachten. Aber es wäre nur eine gewagte Metapher. Die Zurückweisung der Präsupposition einer Frage ist eben keine Verneinung der Frage: sie vereitelt den Frageakt (entzieht ihr die eigentliche Berechtigung), aber sie verneint nicht die Frage.

LITERATUR

BOËR, Steven E. / LYCAN, William G. (1976): "The myth of semantic presupposition". Working papers in linguistics 21: 1-90.

BREKLE, Herbert E. (1972): Semantik. Eine Einführung in die sprachwissenschaftliche Bedeutungslehre. München: Fink.

CONTE, Amedeo G. / DI BERNARDO, Giuliano (1977): "Bibliografia". DI BERNARDO, Giuliano (ed.): Logica deontica e semantica. Bologna: Il Mulino: 347-447.

CONTE, Maria-Elisabeth (im Druck): "Simulare". In: Retorica e critica letteraria. Festschrift für Heinrich Lausberg. Bologna: Il Mulino. (Quaderno 2 di "Lingua e stile".)

DUCROT, Oswald (1972): Dire et ne pas dire. Principes de sémantique linguistique. Paris: Hermann.

HAJIČOVÁ, Eva (1974):"Meaning, presupposition, and allegation." Prager Autorengruppe: Functional generative grammar in Prague. Kronberg (Taunus): Scriptor: 160-172.

KARTTUNEN, Lauri (1970): "On the semantics of complement sentences". Papers from the sixth regional Meeting. Chicago: Chicago Linguistic Society: 328-339.

--- (1971a): "Implicative verbs". Language 47: 340-358.

--- (1971b): "The logic of English predicate complement constructions". Bloomington: Indiana University Linguistics Club (vervielfältigt). Franz. Übersetzung von Maurice Borel: "La logique des constructions anglaises à complément prédicatif". Langages 8 (1973), 30: 56-80.

--- (1973):"Presuppositions of compound sentences". Linguistic Inquiry 4: 169-193.

KIPARSKY, Paul / KIPARSKY, Carol (1970): "Fact". BIERWISCH, Manfred / HEIDOLPH, Karl Erich (eds.): Progress in linguistics. The Hague: Mouton: 143-173.

KÖRNER, Karl-Hermann (1977): Einführung in das semantische Studium des Französischen. Darmstadt: Wissenschaftliche Buchgesellschaft.

LYONS, John (1977): Semantics. London etc.: Cambridge University Press (in zwei Bänden).

MØRDRUP, Ole (1975): "Présuppositions, implications et verbes français". Revue romane 10: 125-157.

PETŐFI, János S. / KAYSER, Hermann (1976): "Sprechhandlungen und semantische Interpretation (Die Rolle der performativen und weltkonstitutiven Ausdrücke in der Textinterpretation)". Universität Bielefeld (Arbeitspapier).
Erscheint in: MEYER-HERMANN, Reinhard (ed.): Sprechen - Handeln - Interaktion. Tübingen: Niemeyer, 1978.

POLDAUF, Ivan (1972): "Factive, implicative, evaluative predicates". Philologica pragensia 15: 65-92.

PRANTL, Carl (1855): Geschichte der Logik im Abendlande. Leipzig: S. Hirzel, I. Band. (Nachdruck: Graz: Akademische Druck- und Verlagsanstalt, 1955.)

PUSCH, Luise F. (1972): Die Substantivierung von Verben mit Satzkomplementen im Englischen und im Deutschen. Frankfurt am Main: Athenäum.

SGALL, Petr / HAJIČOVÁ, Eva / BENEŠOVÁ, Eva (1973): Topic, focus and generative semantics. Kronberg (Taunus): Scriptor.

LA TRANSFORMATIVITE, L'INGRESSIVITE ET L'EGRESSIVITE DANS UNE GRAMMAIRE COMPARATIVE TRANSFORMATIONNELLE DU FRANCAIS ET DE L'ALLEMAND

Jacques François

0. L'analyse qui suit porte sur l'expression du changement d'état en français et en allemand et tente dans cette perspective de fournir une esquisse de réponse aux deux questions suivantes:

(i) A quelles exigences un verbe[1] doit-il satisfaire pour être dit transformatif?

(ii) A quelles exigences un verbe doit-il satisfaire pour être dit ingressif ou égressif?

Pour répondre à ces questions, il me paraît souhaitable

(a) de distinguer clairement le plan du contenu du plan de l'expression;

(b) de rapporter au plan de l'expression les termes "verboïdes transformatifs" et "verboïdes cursifs" empruntés à Fabricius-Hansen (1975: 66-71) et de raffiner sur le plan du contenu la typologie des prédicats d'état et de procès empruntée à Manzotti/Pusch/Schwarze (1975: 29-31)[2] à partir de l'étude sémantique de Lyons (1968: 352-359) sur les adjectifs relatifs et absolus et le travail de Steinitz (1975) sur l'expression des changements d'état.

L'analyse s'achève sur la confrontation des verboïdes français et allemands qui expriment l'état de veille, l'état de sommeil ou la transition de l'un des deux états dans l'autre.

1. Plan du contenu: analyse sémantique de plusieurs sous-classes de prédicats d'état et de procès.

1.1. Considérons à titre d'exemples les phrases allemandes (1-12):

(1) Lise ist jetzt dick.
(2) Lise ist jetzt dünn.
(3) Lise hat jetzt zugenommen.
(4) Lise ist jetzt dicker geworden.
(5) Lise hat jetzt abgenommen.
(6) Lise ist jetzt dünner geworden.
(7) Lise ist jetzt dick geworden.
(8) Lise ist jetzt dünn geworden.
(9) Lise ist jetzt wach.
(10) Lise schläft jetzt.
(11) Lise wacht jetzt auf.
(12) Lise schläft jetzt ein.

Il est clair en première approximation que les phrases (1, 2, 9, 10) expriment des états, tandis que les phrases (3-8, 11, 12) expriment des procès. Toutefois l'examen des prédicats d'état montre (cf. Bartsch/Vennemann 1972: 59-60) que dick et dünn sont deux adjectifs relatifs antonymes qui situent chacun un certain état de corpulence par rapport soit à la norme dick, soit à la norme dünn[3], tandis que schlafen et wach sein expriment des états qui ne peuvent pas se présenter à différents degrés et n'impliquent aucune référence à une norme. Les conditions de vérité (1a, 2a) de (1, 2) comportent donc la référence à une norme (dick ou dünn), à l'inverse des conditions de vérité (9a, 10a) de (9, 10):

(1a) : (1) est vrai si et seulement si (ssi) Lise pèse (m) kg au moment de l'énonciation, (m) étant supérieur ou égal à la norme dick.

(2a) : (2) est vrai ssi Lise pèse (n) kg au moment de l'énonciation, (n) étant inférieur ou égal à la norme dünn.

(9a) : (9) est vrai ssi l'état wach s'applique à Lise au moment de l'énonciation.

(10a) : (10) est vrai ssi l'état schlafen s'applique à Lise au moment de l'énonciation.

De ce fait, je propose de désigner dick sein et dünn sein comme "prédicats d'état graduable normé". A l'inverse schlafen et wach sein sont des "prédicats d'état non-graduable".

Les mêmes distinctions s'appliquent aux prédicats de procès (cf. Steinitz 1975: 14-21). Les phrases (3, 4) et (5,6) qui expriment respectivement les mêmes processus[4] de grossissement et d'amaigrissement n'impliquent pas que l'une des deux normes dick et dünn ait été atteinte ou franchie contrairement aux phrases (7, 8). Il est en effet tout à fait concevable que Lise ait certes grossi des derniers temps, mais qu'elle ait été initialement si maigre qu'elle n'a pas franchi pour autant la norme dick, ou inversement qu'elle ait maigri, mais qu'elle ait été initialement si grosse qu'elle n'a pas franchi la norme dünn. Les conditions de vérité (7a, 8a) de (7, 8) comportent donc la référence aux normes dick et dünn à l'inverse des conditions de vérité (3a, 5a) de (3, 6).

(3a) : (3,4) est vrai ssi Lise pèse (m) kg à l'instant (t_i) et (n) kg à l'instant de l'énonciation (t_{jetzt}), (t_i) étant antérieur à (t_{jetzt}) et (m) inférieur à (n).

(5a) : (5,6) est vrai ssi Lise pèse (m) kg à l'instant (t_i) et (n) kg à l'instant (t_{jetzt}), (t_i) étant antérieur à (t_{jetzt}) et (m) supérieur à (n).

(7a) : (7) est vrai ssi Lise pèse (m) kg à l'instant (t_i) et (n) kg à l'instant (t_{jetzt}), (t_i) étant antérieur à (t_{jetzt}), (m) inférieur et (n) supérieur ou égal à la norme dick.

(8a) : (8) est vrai ssi Lise pèse (m) kg à l'instant (t_i) et (n) kg à l'instant (t_{jetzt}), (t_i) étant antérieur à (t_{jetzt}), (m) supérieur et (n) inférieur ou égal à la norme dünn.

Je propose donc de désigner dicker werden, zunehmen, dünner werden et abnehmen comme "prédicats de procès désignant un changement de degré non normé" et dick/dünn werden comme "prédicats de procès désignant un changement de degré normé". Seules les phrases (11, 12) désignent un changement d'état. Dans les conditions de vérité (11a,

12a) des phrases (11, 12) l'état initial est provisoirement désigné par *nicht-wach* (11a) et *nicht-schlafen* (12a):

(11a) : (11) est vrai ssi à l'instant (t_i) l'état *nicht-wach* et à l'instant (t_k) l'état *wach* s'appliquent à Lise, (t_i) étant immédiatement antérieur et (t_k) immédiatement postérieur à (t_{jetzt}).

(12a) : (12) est vrai ssi à l'instant (t_i) l'état *nicht-schlafen* et à l'instant (t_k) l'état *schlafen* s'appliquent à Lise, (t_i) étant immédiatement antérieur et (t_k) immédiatement postérieur à (t_{jetzt}).

Je propose donc de désigner *wach werden*, *erwachen*, *aufwachen* et *einschlafen* comme des "prédicats de procès désignant un changement d'état".[5] Le tableau (T1) résume cette sous-classification:

(T1) Sous-classification des prédicats d'état et de procès

Préd. d'état	graduable normé		(1,2)
	non graduable		(9,10)
Préd. de procès	changement de degré	non normé	(3,6)
		normé	(7,8)
	changement d'état		(11,12)

1.2. Le propos de mon analyse étant l'expression du changement d'état, je laisserai désormais de côté les prédicats d'état graduable et les prédicats de procès désignant un changement de degré et je limiterai mon étude aux prédicats d'état non graduable et aux prédicats de procès désignant un changement d'état. Pour cela j'introduis les conventions d'écriture suivantes: l'assignation à Lise du prédicat d'état *schlafen* (ou *wach sein*) au moment de l'énonciation est noté $S(\text{Lise},t_{jetzt})$, $W(\text{Lise},t_{jetzt})$. L'assignation à Lise du prédicat de procès "transition de *nicht-wach* à *wach*" (ou "transition de *nicht-schlafen* à *schlafen*") à l'instant de l'énonciation est notée respectivement $\Delta_{\bar{W},W}(\text{Lise},t_{jetzt})$ et $\Delta_{\bar{S},S}(\text{Lise},t_{jetzt})$.

Les prédicats d'état $W(Lise,t_i)$ et $S(Lise,t_i)$ entretiennent une relation de complémentarité (cf. Lyons 1968: 352-359 et Schwarze 1975: 79-80) notée $\succ\!\!-\!\!\prec$ qui se définit ainsi (13):

(13) $\forall t\ W(Lise,t) \succ\!\!-\!\!\prec S(Lise,t) =_{df}$

$$W(Lise,t) \supset \bar{S}(Lise,t) \wedge \bar{W}(Lise,t) \supset S(Lise,t) \wedge$$
$$S(Lise,t) \supset \bar{W}(Lise,t) \wedge \bar{S}(Lise,t) \supset W(Lise,t)$$

De ce fait, nous pouvons substituer schlafen à nicht-wach et wach à nicht-schlafen dans les conditions de vérité de (11, 12):

(11b) : (11) est vrai ssi à l'instant (t_i) l'état schlafen et à l'instant (t_k) l'état wach s'appliquent à Lise, etc.

(12b) : (12) est vrai ssi à l'instant (t_i) l'état wach et à l'instant (t_k) l'état schlafen s'appliquent à Lise, etc.

Il en résulte que les notations $\Delta_{\bar{W},W}(Lise,t_i)$ et $\Delta_{S,W}(Lise,t_i)$ sont équivalentes ainsi que les notations $\Delta_{\bar{S},S}(Lise,t_i)$ et $\Delta_{W,S}(Lise,t_i)$. Le tableau (T2) illustre les relations de sens entre les phrases (9-12), l'adverbe de temps jetzt désignant dans chaque phrase des moments d'énonciation différents.

(T2) Relationsde sens entre les phrases (9-12)

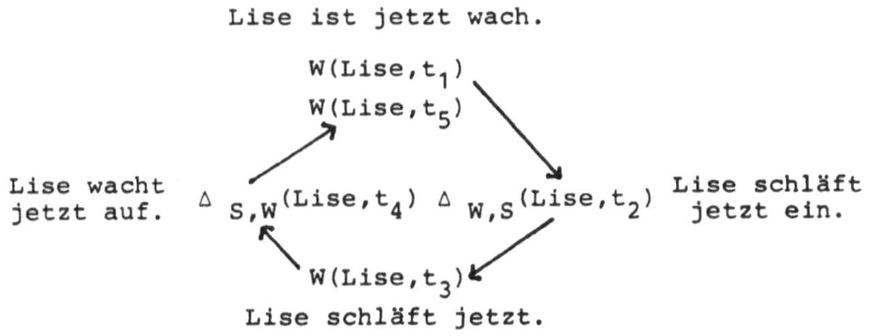

Dans d'autres cas cependant, un prédicat d'état entretient seulement une relation d'opposition disjonctive avec un nombre fini d'autres prédicats d'état (cf. Fabricius-Hansen 1975: 80-83 et Wunderlich 1970: 144). Ainsi, pour les prédicats d'état correspondant aux verbes de position allemands liegen, stehen, sitzen, hängen et hocken ces relations d'apposition disjonctive seraient:

(15) $\forall t\ \overline{Li}(a,t) \equiv St(a,t)\ v\ Si(a,t)\ v\ Hä(a,t)\ v\ Ho(a,t)$

$\forall t\ \overline{St}(a,t) \equiv Li(a,t)\ v\ Si(a,t)\ v\ Hä(a,t)\ v\ Ho(a,t)$, etc.

Si l'on cherche à représenter le contenu de (16)

(16) Karl steht auf.

on voit que seule la formulation $\Delta_{\overline{St},St}(Karl,t_{jetzt})$ convient et que seule une information contextuelle complémentaire du genre de (17)

(17) Bis jetzt saß Karl.

permet de substituer $\Delta_{Si,St}(Karl,t_{jetzt})$ à $\Delta_{\overline{St},St}(Karl,t_{jetzt})$.

Dans la suite de l'analyse je laisserai de côté les prédicats de procès du type aufstehen pour me limiter aux prédicats du type einschlafen, erwachen qui expriment la transition entre deux états complémentaires. D'autre part je ne mentionnerai plus la localisation temporelle des états et des procès qui n'intervient plus directement dans la discussion.

2. Plan de l'expression: analyse morphosyntaxique des verboïdes cursifs et transformatifs.

2.1. Les deux classes de verboïdes qui entrent ici en ligne de compte, les verboïdes cursifs et les verboïdes transformatifs (respectivement corrélatifs sur le plan de l'expression des prédicats d'état non-graduable et des prédicats de procès exprimant un changement d'état sur le plan du contenu) se réalisent sous des formes très diverses:

2.1.1. Réalisation superficielle des verboïdes (V^*) cursifs[6]

(a) VCurs : (18) Die Rose blüht.

(19) Blüh(a) ⁓>

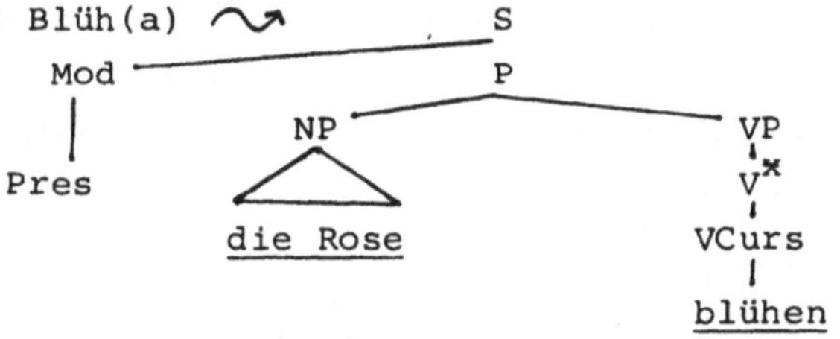

(b) AuxCurs + Adj : (20) Lise ist wach.

(21) Wach(Lise) ⁓>

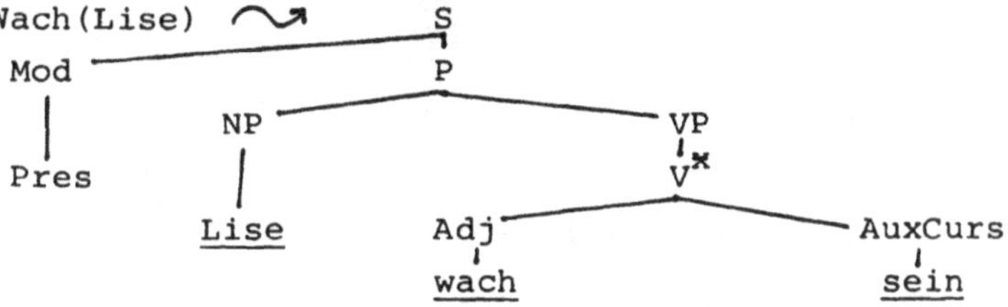

(c) AuxCurs + [Prep+NP] : (22) Die Rose steht in (voller) Blüte.

(23) Blüh(a) ⁓>

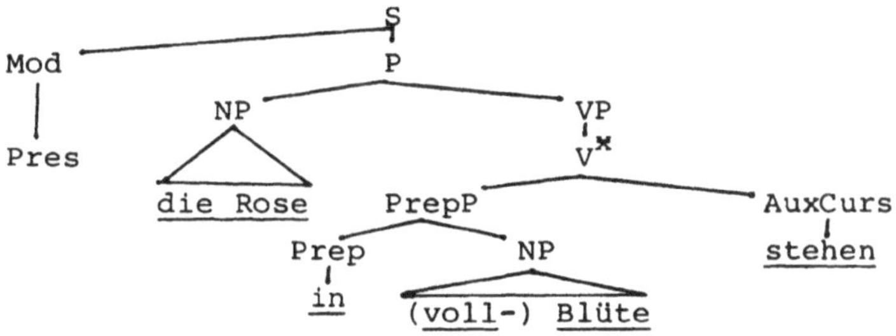

2.1.2. Réalisation superficielle des verboïdes transformatifs ingressifs

La thèse centrale de Fabricius-Hansen est qu'il n'existe pas de "verbes ingressifs" et de "verbes égressifs", mais seulement des relations d'ingressivité et d'égressivité entre des verboïdes cursifs et des verboïdes transformatifs. Cette thèse est cohérente dans la mesure où les notions d'ingressivité et d'égressivité se rapportent au plan du contenu. En revanche si, comme je le propose ici, on rapporte les concepts d'ingressivité et d'égressivité au plan de l'expression, les verboïdes transformatifs se subdivisent naturellement en verboïdes ingressifs, égressifs et transgressifs. Les <u>verboïdes ingressifs</u> sont des verboïdes transformatifs obtenus par dérivation ou auxiliation à partir d'un verbe cursif (V^r), d'un adjectif (Adj^r) ou d'un radical verbal (RV^r) exprimant l'état résultant du procès:

(a) AuxIng + V^r: (24) Pierre commence à délirer.

(25)
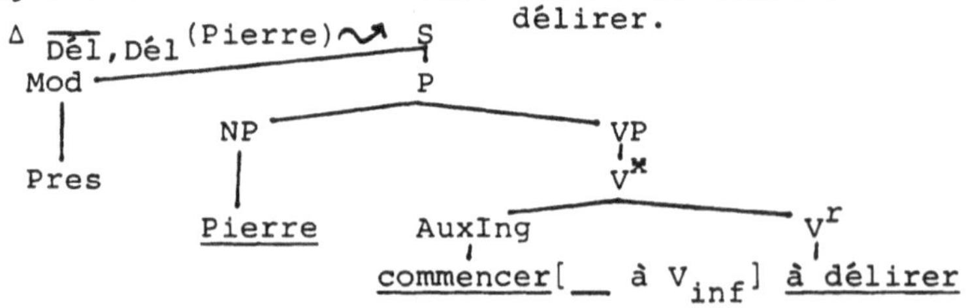

(b) AuxIng +(Det)+ $[V^r]_N$: (26) Pierre prend peur.

(27)
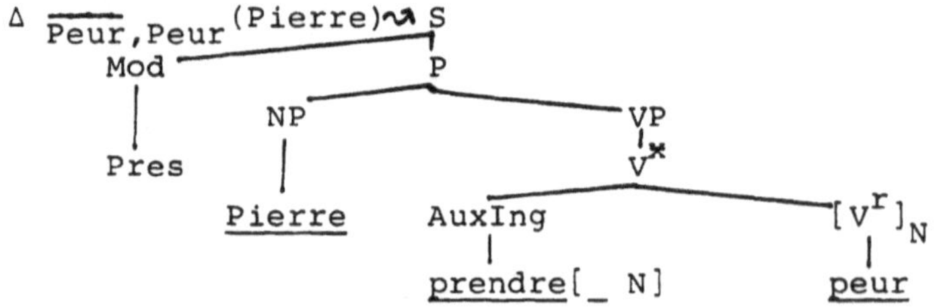

(c) AuxIng +[Prep +(Det)+ $[V^r]_N$] : (28) Otto gerät in Wut

(29)
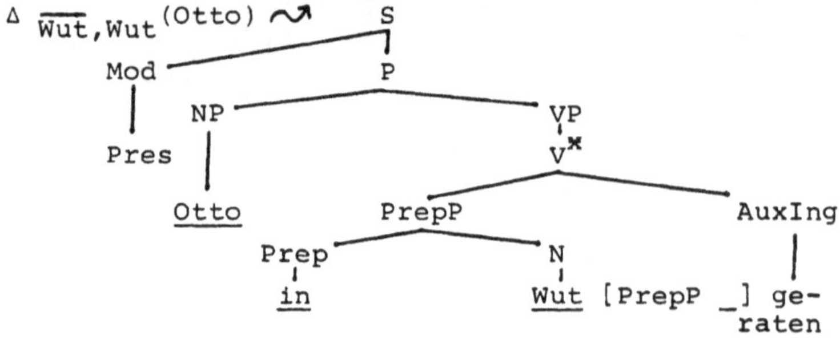

(d) AuxIng + Adj^r : (30) Petra devient folle.

(31)
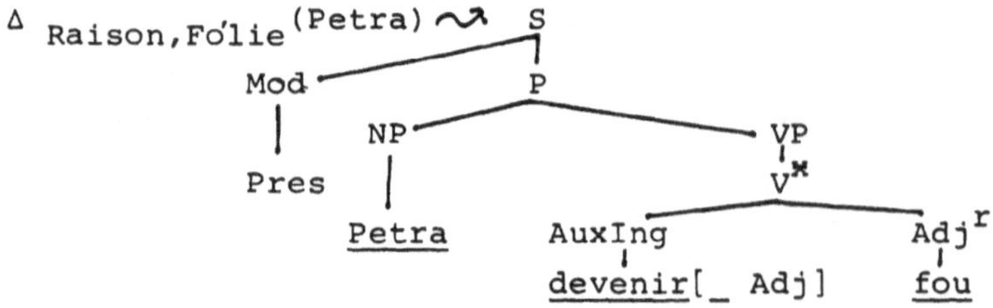

(e) PrevIng + V^r : (32) Lise schläft ein.

(33) $\Delta_{W,S}$(Lise) ↝

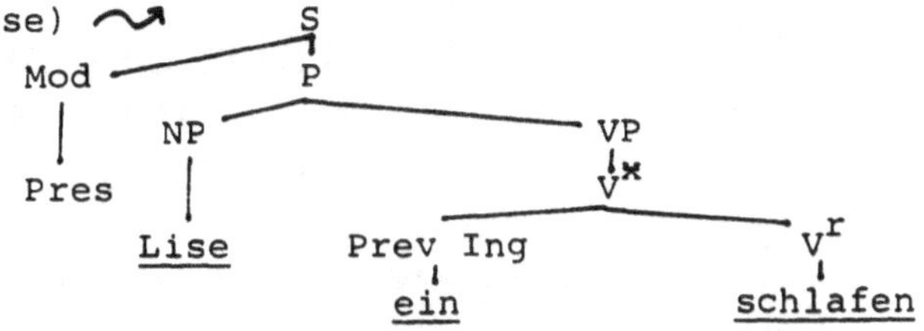

(f) $_V$[PrevIng + RV^r] : (34) Lise erwacht.

(35) $\Delta_{S,W}$(Lise) ↝

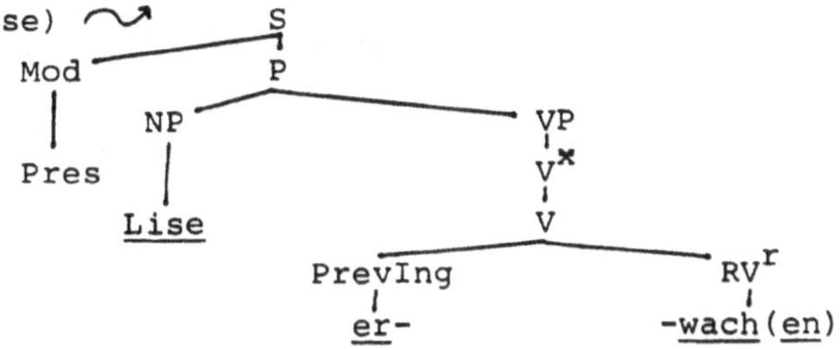

(g) ProRef + $_V$[PrevIng + RV^r]: (36) Lise s'éveille.

(37) $\Delta_{S,W}$(Lise) ↝

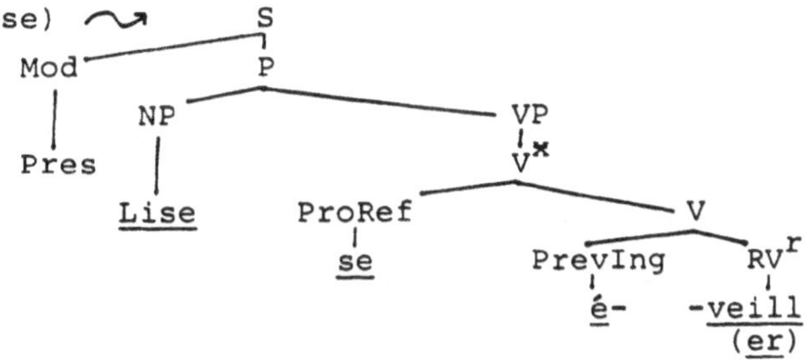

2.1.3. Réalisation superficielle des verboïdes égressifs

Les <u>verboïdes égressifs</u> sont des verboïdes transformatifs obtenus par dérivation ou auxiliation à partir d'un verbe cursif (V^i), d'un adjectif (Adj^i) ou d'un radical verbal (RV^i) exprimant l'état initial du procès:

(a) AuxEgr + V^i : (38) Pierre cesse de délirer. (cf.24,25)

(b) AuxEgr + (det)+$[V^i]_N$: (39) Petra perd la raison. (cf. 26,27)

(c) AuxEgr + Prep +(Det)+$[V^i]_N$:
(40) Lise sort du sommeil. (cf.28,29)

(d) PrevEgr + V^i : (41) Das Lied klingt aus. (cf.32,33)

(e) $_{V}[PrevEgr + RV^{i}]$: (42) Die Rose verblüht. (cf.34,35)

2.1.4. Les verboïdes transgressifs sont des verboïdes transformatifs qui se réalisent sous la forme d'un lexème verbal non dérivé, ou dérivé à partir d'un lexème exprimant la cause matérielle du procès. Schmelzen fait partie de la première catégorie (par opposition aux ingressifs zerfliessen et sich verflüssigen, cf. 43), verhungern et verdursten de la seconde.

(43) $\Delta_{\overline{Flüssig},Flüssig}(a)$ ↝

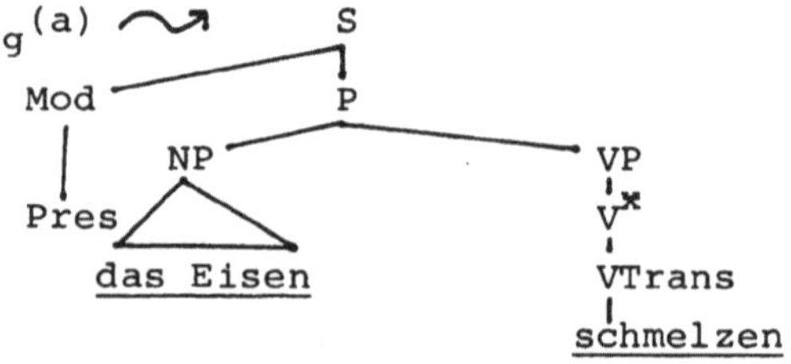

2.2. Il ressort de la confrontation des phrases et des représentations (25-43) que le français et l'allemand disposent d'un répertoire important (et dans une large mesure parallèle) de formes ingressives, égressives et transgressives pour exprimer le changement d'état. Lorsque l'état initial et l'état résultant entretiennent une relation de complémentarité les phrases comportant un verboïde ingressif et les phrases comportant un verboïde égressif entretiennent une relation d'équivalence sémantique (si tous les autres éléments des deux phrases sont identiques). Ainsi (39) Petra perd la raison et (31) Petra devient folle expriment à l'aide d'une forme soit égressive (39), soit ingressive (31) la même transition $\Delta_{Raison,Folie}$(Petra). Il en est de même pour (42) Die Rose verblüht qui exprime la transition $\Delta_{Blüh,Welk}(a)$ à l'aide d'une forme égressive et (44) Die Rose wird welk et (45) Die Rose verwelkt qui expriment le même changement d'état à l'aide d'une forme ingressive.

2.3. Si nous essayons de porter sur le tableau (T2) les verboïdes cursifs et transformatifs français et allemands qui expriment l'état de veille, l'état de sommeil et la transition entre ces deux états, nous constatons (cf. T3) que chacune des deux langues privilégie pour une part des réalisations différentes des verboïdes cursifs et transformatifs. Ainsi il n'y a pas en français d'équivalent isomorphe de *wach werden* ni même de *wach sein*, car *être éveillé* (comme *être endormi*) est un verbe transformatif ingressif à "l'aspect accompli"[7]. En outre la préfixation joue en allemand un rôle plus systématique dans l'expression de la transformativité. Toutefois les deux langues convergent pour l'expression du changement d'état $\Delta_{S,W}$ et $\Delta_{W,S}$ dans le choix de verboïdes ingressifs au détriment des verboïdes égressifs absents de (T3) à l'exception de la tournure recherchée *sortir du sommeil*.

(T3)

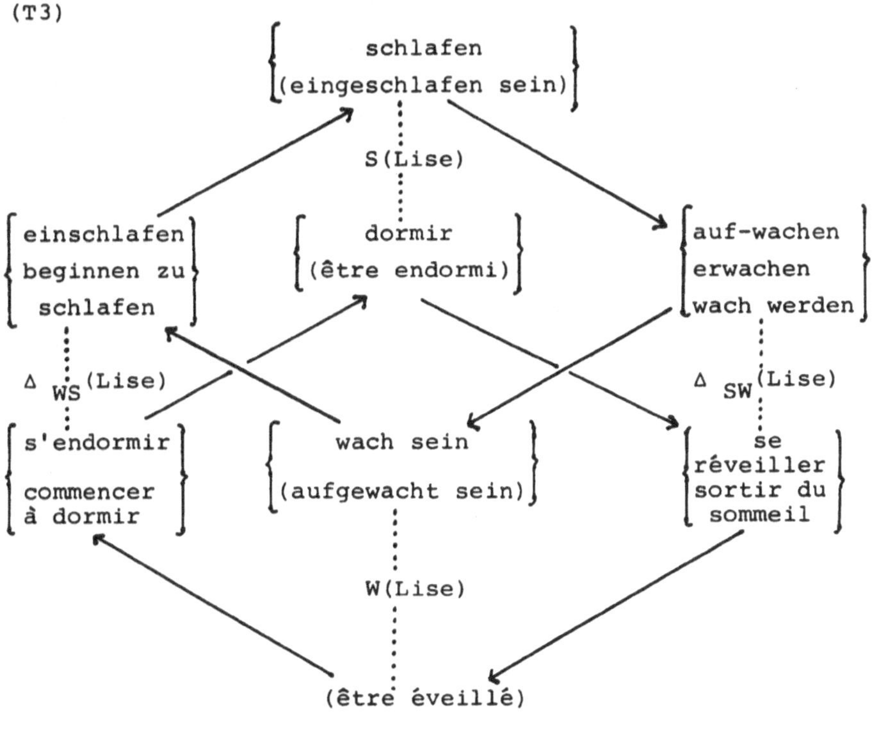

3. Pour conclure, je soulignerai certains problèmes majeurs qui restent à résoudre dans l'analyse du changement d'état sur le plan du contenu et sur le plan de l'expres-

sion. Sur le plan du contenu il reste à définir des tests permettant de distinguer dans tous les cas les prédicats de procès désignant un changement de degré et ceux désignant un changement d'état. Il importe également de distinguer les prédicats de procès des prédicats d'action réfléchie: par ex. sich beruhigen est interprétable soit comme ruhig werden soit comme sich zur Ruhe zwingen. Dans le premier cas nous avons affaire à un procès, dans le second à une action réfléchie et il reste à déterminer si l'opposition de sens doit se refléter sur le plan du lexique ou sur celui de la représentation syntaxique.

Sur le plan de l'expression il reste entre autres à spécifier la façon dont certaines réalisations morphologiques des verboïdes transformatifs sont en distribution complémentaire, par ex. devenir [_ Adj], commencer [_ à V_{inf}], entrer [_ en N], prendre [_ N]. Il importe également de rendre compte des modalités de l'ingressivité et de l'égressivité qui apparaissent dans les définitions lexicographiques. Ainsi le Deutsches Wörterbuch de G. Wahrig accorde une place privilégiée aux modalités allmählich (verklingen $=_{df}$ allmählich aufhören zu klingen) et langsam (verglimmen $=_{df}$ langsam aufhören zu glimmen). Il reste enfin à rendre compte de la place occupée dans chacune des deux langues par les réalisations ingressives, égressives et transgressives de la transformativité.

NOTES

1 La transformativité et la cursivité sont, selon Fabricius-Hansen (1975: 87) des propriétés des "verboïdes" (Verbale) qui se subdivisent superficiellement en lexèmes verbaux (einfache Verben), syntagmes attributifs (Prädikativfügungen), syntagmes verbaux fonctionnels (Funktionsverbfügungen) et séquences verbales subordinatives (subordinative Verbalketten). Nous reviendrons sur cette classification au § 2.1.

2 Les problèmes abordés dans Manzotti/Pusch/Schwarze 1975

sont développés par les mêmes auteurs dans "Kieler Mehrsprachengrammatik" (à paraître). Toutefois le traitement des prédicats d'état et de procès y demeure identique.

3 Les adjectifs relatifs désignent soit des propriétés, soit des états. Lise ist dick peut donc s'interpréter soit dans le sens Lise ist von Natur aus dick soit dans le sens Lise ist jetzt dick. C'est cette interprétation stative qui s'applique dans la suite de l'analyse aux phrases (1,2) du fait de la présence de jetzt.

4 Quoique les phrases (3,4) expriment le même processus de prise de poids, il semble que l'emploi de dicker werden soit soumis à des restrictions (par ex. à la seconde personne) auxquelles n'est pas soumis zunehmen.

5 Dans Manzotti/Pusch/Schwarze (1975: 30) la sous-classification des prédicats de procès entre "prédicats qui désignent un changement d'état absolu et prédicats qui désignent un changement graduel" est rejetée "parce que l'intension des prédicats de procès pris isolément permet presque toujours l'une ou l'autre possibilité." En exemple sont cités Das Licht ging plötzlich/langsam aus et Er stirbt seit drei Monaten. Cependant dans les exemples cités le changement d'état ne fait pas de doute, seules diffèrent la progressivité ou la soudaineté du procès, alors que dans les phrases (3-6) nous avons affaire à un procès sans changement d'état, c'est-à-dire à une progression dans les limites d'un même "état de corpulence".

6 Dans les représentations qui suivent le symbole $\rightsquigarrow$ emprunté à Bartsch/Vennemann (1972) désigne la relation entre une structure sémantique et une structure syntaxique proche de la surface. Les représentations syntaxiques employées visent uniquement à montrer la place du symbole V^{*} sur l'arbre et les différentes sortes de décomposition des verboïdes.

7 (Etre) éveillé peut être considéré soit comme adjectif (absolu) soit comme participe passé du transformatif s'éveiller. En allemand, du fait de l'existence du couple wach sein/aufgewacht sein, il y a une relation d'implication entre deux prédicats d'état: aufgewacht (Lise) $\supset$ (unmittelbar danach) wach (Lise). En français, l'absence d'équivalent isomorphe de wach sein a pour effet la neutralisation en surface de cette relation si bien que être éveillé et wach sein sont fonctionnellement équivalents.

BIBLIOGRAPHIE

BARTSCH, Renate & VENNEMANN, Theo (1972): Semantic structures. A study in the relation between semantics and syntax. Frankfurt/Main: Athenäum.

FABRICIUS-HANSEN, Cathrine (1975): Transformative, intransformative und kursive Verben. Tübingen: Niemeyer.

LYONS, John (1968): Introduction to theoretical linguisties. Cambridge: Cambridge University Press. Cité d'après l'édition française. Paris: Larousse (1970).

MANZOTTI, E. & PUSCH, L.F. & SCHWARZE, Ch. (1975): Sorten von Prädikaten und Wohlgeformtheitsbedingungen für eine Semantiksprache. Zeitschrift für germanistische Linguistik 3.1. Berlin: de Gruyter.

SCHWARZE, Christoph (1975): Einführung in die Sprachwissenschaft, mit Beispielen aus dem Französischen und dem Deutschen. Kronberg: Scriptor, Monographien Linguistik und Kommunikationswissenschaft 21.

STEINITZ, Renate (1975): Sind alle Inchoativa inchoativ? Linguistische Studien, Reihe A: Arbeitsberichte, 18. Berlin: Akademie der Wissenschaften der DDR, Zentralinstitut für Sprachwissenschaft.

WAHRIG, Gerhard (1968): Deutsches Wörterbuch. Gütersloh: Bertelsmann.

WUNDERLICH, Dieter (1970): Tempus und Zeitreferenz im Deutschen. München: Hueber, Linguistische Reihe 5.

KONTEXT UND KENNZEICHNUNG

Manfred Pinkal

1. Hypothesen über sprachliche Sachverhalte hören sich oft recht trivial an, wenn man von einer exakten Formulierung absieht und sich auf informelle Umschreibungen beschränkt. Die formale Beschreibung von Sprache - insbesondere die Beschreibung sprachlicher Bedeutungen - setzt jedoch in vielen Fällen einen komplexen Apparat voraus, der sich nicht auf wenigen Seiten sinnvoll darstellen läßt. Auch auf die Gefahr hin, daß meine Ausführungen trivial klingen, muß ich hier deshalb weitgehend auf eine formale Analyse der Kennzeichnungsproblematik verzichten. Stattdessen werde ich versuchen, die intuitiven Probleme zu charakterisieren, die sich bei den gängigen Theorien zu definiten Deskriptionen ergeben, und anzudeuten, wie mein Ansatz zur Interpretation definiter Nominalphrasen einige dieser Probleme bewältigt[1].

Zur Einordnung der Analyse nur soviel: Der verwendete syntaktische und semantische Rahmen ist eine erweiterte und modifizierte Fassung des von M.J. Cresswell entwickelten λ-kategorialen Apparates[2]. Zur Behandlung des zentralen Problems semantischer Kontextabhängigkeit kombiniere ich, grob gesagt, den in LEWIS (1972) entwickelten coordinate approach und eine Variante der supervaluation-Theorie.[3]

2. Zur Eingrenzung des untersuchten Phänomens: Nicht zur Diskussion steht im folgenden der generische Gebrauch des bestimmten Artikels wie in (1):

(1) Die Tasse ist ein Trinkgefäß

Ebensowenig finden Probleme Berücksichtigung, die sich im Zusammenhang mit Plural-Nominalphrasen oder mit der Verwendung des bestimmten Artikels bei kollektiven, kontinuativen und abstrakten Nomina ergeben. Ich beschränke mich vielmehr auf den paradigmatischen, ontologisch unproblematischen Fall der Kennzeichnung einzelner diskreter physischer Objekte.

Auch bei dieser Einschränkung auf die Standardverwendung lassen sich eine Reihe verschiedener semantisch-pragmatischer Kennzeichnungstypen unterscheiden. HAWKINS (1977) etwa differenziert zwischen acht Typen, die ihrerseits in zahlreiche Subtypen zerfallen. Ich lasse es bei einer - allerdings fundamentalen - Unterscheidung zweier Klassen bewenden. Zur ersten Klasse gehören Kennzeichnungen wie der König von Frankreich, der König, der Direktor, der Sieger, der Sieger im Weitsprung, der Sieger bei der Olympiade 76, der Sieger im Weitsprung bei der Olympiade 76, der Ehemann, der Ehemann von Annegret, die Fünfte von Beethoven; mit Einschränkungen auch die Haustür, der Verteilerkopf, der Blinddarm. Vom zweiten Typ sind die Tasse, der Baum, der rote Ball, das Haus mit der Birke im Vorgarten, der grüne 2CV.[4]

Semantisch läßt sich der Unterschied zwischen den beiden Klassen grob so bestimmen, daß Kennzeichnungen des ersten Typs Funktionsausdrücke enthalten - entweder explizit oder in verkürzter Form -, während es sich im zweiten Fall um 'normale' Prädikate handelt[5]. Dem entspricht ein tiefgreifender Unterschied in der Art und Weise, wie die Bedeutung der gesamten Kennzeichnung vom Äußerungskontext abhängt.

(2) Der König ist kahlköpfig

Bei Kennzeichnungen vom Typ I muß, wie aus dem Beispiel (2) ersichtlich, vom Äußerungskontext nicht direkt ein Referent spezifiziert werden, sondern ein Argument für den entsprechenden Funktionsausdruck (König von). Ganz anders verhält es sich bei Ausdrücken des Typs II:

(3) Gib mir die Tasse

(4) Das Haus hat einen neuen Anstrich nötig

Äußerungen von Sätzen wie (3) oder (4) sind nur dann erfolgreich, wenn für den Hörer der referierte Gegenstand selbst durch die Kommunikationssituation eindeutig spezifiziert ist. Diesem Unterschied entspricht unter anderem die Tatsache, daß für NPs des zweiten Typs im Gegensatz zum ersten ein de dicto-Gebrauch nicht möglich ist; man vergleiche (5) und (6):

(5) der jeweilige/jetzige König

(6) +die jeweilige/jetzige Tasse

Auf die intuitiv weniger problematische Semantik von Ausdrücken des ersten Typs kann ich hier nicht näher eingehen. Ich beschränke mich im folgenden auf die Diskussion der zweiten Klasse von Kennzeichnungen. Deren Interpretation kann sich - wie aus der obigen Gegenüberstellung und der folgenden Betrachtung einiger Interpretationsversuche ersichtlich - weder direkt an die Analyse des ersten Typs anlehnen noch auf den ersten Typ übertragen werden.

3. Gemäß der klassischen Russellschen Interpretation definiter Deskriptionen, die unter anderem auch von Montague übernommen wurde[6], bezeichnet Satz (7) einen Sachverhalt, der ziemlich genau durch (8) umschrieben wird.

(7) Die Tasse hat einen Sprung

(8) Es gibt einen und nur einen Gegenstand, der eine Tasse ist, und dieser Gegenstand hat einen Sprung

Die Analyse ist aus zwei Gründen unzureichend: Erstens wird in (7) der erste Teilsatz von (8) nicht behauptet, sondern vorausgesetzt[7]. Diesem Einwand kann man gerecht werden, indem man die Existenz genau einer Tasse nicht als Teil der wahrheitsfunktional relevanten Bedeutung des Satzes auffaßt, sondern als Präsupposition behandelt[8]. Damit ist für die intuitive Adäquatheit der Analyse jedoch nicht viel gewonnen; denn - dies ist der zweite Einwand - (7) ist durchaus nicht nur in dem Fall interpretierbar, daß es genau eine Tasse gibt - was unmittelbar daraus hervorgeht, daß in diesem unserem mit zahlreichen Tassen versehenen Universum Satz (7) in vielen Situationen sinnvoll verwendet werden kann [9].

Weiter führt die Formulierung, daß (7) nur dann interpretierbar ist, wenn der Äußerungskontext die Existenz- und Eindeutigkeitspräsupposition erfüllt, wenn also in der Äußerungssituation 'genau eine Tasse vorkommt'. Diese Formulierung, die sinngemäß den Ausführungen in KARTTUNEN 1974 entspricht, vermeidet einerseits die starke Vorannahme über den Zustand der Welt; andererseits berücksichtigt sie die Tatsache, daß die Referenz definiter Deskriptionen situationsabhängig ist. Sie erklärt leider nicht, wieso eine sinnvolle Verwendung von (7) auch an einem für mehrere Personen gedeckten Frühstückstisch oder am Fließband in einer Porzellanmanufaktur möglich ist.

Cresswells Behandlung des bestimmten Artikels versucht, diesem Phänomen gerecht zu werden. Ich kann seine schon recht komplizierte Analyse hier ebenfalls nur knapp skizzieren[10]. Zum Äußerungskontext in Cresswells Verständnis gehört grundsätzlich alles: informell umschrieben, ist 'Kontext' etwa 'die ganze Welt relativ zu einem spezifischen Äußerungsereignis mit Ausnahme des geäußerten Satzes'. Zum Kontext gehören somit grundsätzlich alle Häuser, alle Bäume, alle Tassen. Die situationsbezogene Formulierung der Eindeutigkeitspräsupposition fiele bei Zugrundelegung dieser Auffassung von Kontext also mit der faktischen Einmaligkeitsforderung der voraufgehenden Formulierungen zusammen. Nun haben Kontexte nach Cresswell aber unter anderen Eigenschaften die, bestimmte Mengen von Individuen zu spezifizieren, die als potentielle Referenten für geäußerte definite Nominalphrasen in Frage kommen ('e-specifying properties'). Eine definite Deskription der/die/das α ist nach Cresswell in genau den Kontexten interpretierbar, deren Menge möglicher Referenten ein und nur ein α enthält. Am Beispiel: Satz (7) kann in allen Kontexten sinnvoll geäußert werden, die - neben anderen Individuen wie Häusern, Bäumen, Hunden, Büchern genau e i n e Tasse spezifizieren (etwa indem der Sprecher darauf zeigt oder der Adressat der Äußerung gerade nach ihr greift oder direkt vorher Satz (9) geäußert wurde).

(9) Nimm doch einfach die Tasse, die vor dir steht

Cresswells Interpretation des bestimmten Artikels führt offenbar zu einer Analyse definiter Deskriptionen, die wesentlich adäquater ist als ihre Vorläufer, weil sie wesentlich schwächere Voraussetzungen macht: Es braucht nicht, um im Beispiel zu bleiben, genau eine Tasse zu e x i s t i e r e n , es braucht auch nicht genau eine Tasse in der Äußerungssituation v o r z u - k o m m e n , es muß lediglich die Bedingung erfüllt sein, daß vom Äußerungskontext genau eine Tasse als möglicher Referent s p e z i f i z i e r t wird. Leider ist auch diese Bedingung noch zu stark. Im selben Kontext geäußert, können die Subjekts-NPs der Sätze (10) - (12) sich auf verschiedene Gegenstände beziehen und, jede für sich, dennoch problemlos interpretierbar sein:

(10) Die Tasse hat einen Sprung

(11) Die Teetasse hat einen Sprung

(12) Die blaue Teetasse hat einen Sprung

(10) etwa könnte sich auf die Tasse beziehen, die der Sprecher gerade in der Hand hält, (11) auf ein am gestrigen Abend benutztes Trinkgefäß, (12) auf irgendeine Tasse in einem Wandregal im Nebenzimmer. Zwischen den in (10) - (12) verwendeten Prädikaten besteht offenbar die Inklusionsrelation: eine blaue Teetasse ist immer eine Teetasse, eine Teetasse immer eine Tasse.

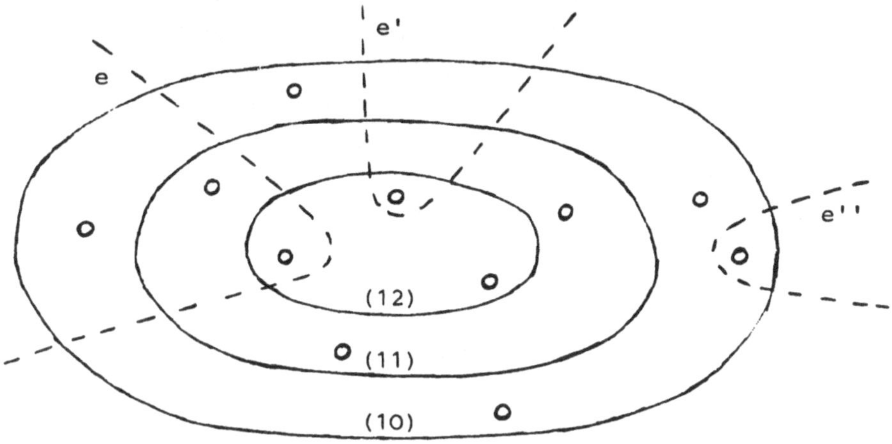

Wenn nun - gemäß Cresswells Analyse - die vom Kontext spezifizierte Menge möglicher Referenten drei verschiedene Gegenstände für die Sätze (10) - (12) vorsieht (Fall e in der Skizze), handelt es sich dabei jedenfalls um drei Tassen, und (10) ist, ebenso wie (11), wegen kontextspezifischer Mehrdeutigkeit nicht interpretierbar. Spezifiziert der Kontext dagegen genau eine Tasse, so ist die Referenz der Kennzeichnungen in (10) - (12) entweder identisch (e': es handelt sich um eine blaue Teetasse - aber dies muß nicht der Fall sein), oder (12) (und ggf. (11)) bleibt in Ermangelung eines Referenten uninterpretiert (e''). Das Phänomen läßt sich also im Rahmen von Cresswells Analyse nicht intuitiv zufriedenstellend behandeln. Die Analyse führt bei allen Sprachen, in denen Hierarchien von Prädikaten vorkommen - und damit wohl bei allen natürlichen Sprachen - zu intuitiv inadäquaten Ergebnissen. Diese partielle Inadäquatheit wiederum scheint mir eng zusammenzuhängen mit der grundsätzlichen intuitiven Problematik eines Kommunikationsmodells, in dem jede Äußerungssituation für jede geäußerte Kennzeichnung ein mögliches Referenzobjekt explizit bereitstellt.

4. Was folgt, ist die Skizze meines Ansatzes zur semantischen Beschreibung definiter Deskriptionen, die die Forderung nach eindeutiger Spezifikation eines Referenten durch den Kontext nicht mehr enthält, oder besser: sie so uminterpretiert, daß das am Beispiel der Sätze (10) - (12) vorgeführte Problem sich von selbst erledigt (zusammen mit einigen weiteren Problemen, auf die ich abschließend eingehen werde).

Die Theorie der semantischen Kontextabhängigkeit, in deren Rahmen ich die Analyse der Kennzeichnungen vornehme, ist zweistufig. Die erste Stufe ist eine konventionelle analytische Semantik mit Kontextabhängigkeit, die in etwa LEWIS 1972 entspricht, aber Lücken in der Wahrheitswertzuweisung zuläßt. Der bestimmte Artikel wird in der ersten Stufe als eine Art erweitertes Personalpronomen definiert[11]; über eine Kontextkoordinate für den 'referierten Gegenstand' ergibt sich, wie beim Pronomen, der kontextspezifische Wert der jeweiligen Nominalphrase. Die Sätze (10) und (13) wären also gleichbedeutend.

(13) Sie hat einen Sprung

Der Unterschied zwischen (10) und (13) besteht darin, daß der Prädikatausdruck der Subjekts-NP in (10) deren Interpretation in allen Kontexten blockiert, in denen der referierte Gegenstand keine Tasse ist.

Soweit ist die Analyse recht unintuitiv: Der Prädikatausdruck in der definiten NP hätte keine weitere Funktion, als in den allermeisten Fällen das Gelingen von Kommunikation zu verhindern Tatsächlich ergibt sich der eigentliche Effekt der Analyse erst in der zweiten, induktiven Stufe der Theorie, die ich wiederum nur andeutungsweise charakterisieren kann.

Will jemand einen in einer normalen Kommunikation geäußerten kontextabhängigen Satz verstehen, steht ihm zu diesem Zweck der vollständige Äußerungskontext in der Regel nicht zur Verfügung; er kennt nur Bruchstücke dieses Kontextes, die insgesamt sein Wissen über den Gesprächszusammenhang ausmachen[12]. Anders formuliert: Der Bewertung einer Äußerung durch ihren Adressaten liegt nicht e i n Kontext zugrunde, sondern die Menge der Kontexte, die mit seiner Kenntnis der Situation verträglich sind. Folglich findet die Bewertung einer Äußerung im Normalfall nicht als Berechnung ihres semantischen Wertes aus vorgegebenen Kontextdaten statt, sondern als Schätzung eines Wertes auf der Grundlage eines Kontextfragments (bzw. der mit diesem Fragment verträglichen Kontextmenge).

Der für die Bewertung definiter Deskriptionen relevante 'referierte Gegenstand' ist vermutlich selten direkt gegeben; er muß aus anderen, für den Hörer greifbaren Zügen der Situation, die mit der Referenzkoordinate korrelieren, erschlossen werden (z.B. Blickrichtung des Sprechers, im letzten Satz referierte Objekte). Während nun bei der Bewertung eines Personalpronomens, wie es in (13) vorkommt, die Schätzung des referierten Gegenstandes auf der Basis der gesamten Kontextmenge erfolgen muß, verhält es sich bei den Subjekts-NPs der Sätze (10) - (12) anders: Entsprechend der Interpretation des bestimmten Artikels in der ersten Stufe scheiden aus der Menge der mit der Situationskenntnis des Hörers verträglichen Kontexte diejenigen aus, in denen der referierte Gegenstand keine Tasse bzw. Teetasse bzw. blaue Teetasse ist. Definite Deskriptionen haben also die Funktion, die Basis für die Schätzung des von ihnen referierten Gegenstandes zu spezifizieren. In inhaltlicher Formulierung: Satz (10) wird bewertet wie Satz (13), nachdem vorher die Situationskenntnis des Hörers um die Information angereichert worden ist, daß es sich beim referierten Objekt um eine Tasse handelt.

5. Die Interpretation leistet das Gewünschte[13]: Durch verschiedene Prädikatausdrücke wird die Basis der Bedeutungsschätzung verschieden stark eingeschränkt. Bei gleicher Ausgangssituation kann für jede der Kennzeichnungen in (10) - (12) also

genau ein Referent ermittelt werden, ohne daß die Referenten identisch zu sein brauchen. Darüber hinaus hat die Analyse einige weitere Vorzüge, von denen ich die wichtigsten abschliessend aufzählen möchte:

(i) Die Eindeutigkeitspräsupposition kann reformuliert und so der Intuition angepaßt werden: Zur Bewertung eines Satzes mit definiten Deskriptionen ist weder die Existenz noch die Spezifikation genau eines referierten Objektes erforderlich, sondern nur hinreichende Eindeutigkeit bei der Referentenschätzung; die Eindeutigkeit wiederum kann sich sowohl aus der Situation wie aus spezifizierenden Prädikatausdrücken ergeben.[14]

(ii) Definite Deskriptionen und (referierende) Personalpronomen können parallel interpretiert und auf denselben Typ von Kontextabhängigkeit zurückgeführt werden.

(iii) Ohne weitere Modifikationen der Analyse lassen sich im Rahmen meiner zweistufigen Theorie der semantischen Kontextabhängigkeit auch solche Deskriptionen intuitiv adäquat behandeln, bei denen der Prädikatausdruck selbst kontextabhängig ist (<u>die große Tasse</u>, <u>die Tasse dort drüben</u>). - Wie dies Zusammenspiel verschiedener Typen von Kontextabhängigkeit in die Schätzung des referierten Gegenstandes eingeht, kann ich allerdings an dieser Stelle nicht mehr ausführen.

ANMERKUNGEN

1 zum genaueren Verständnis des Formalismus s. PINKAL 1977

2 ausführlich dargestellt in CRESSWELL 1973. - Für die Verwendung von Cresswells λ-kategorialer Grammatik spricht deren im Vergleich zur Montague-Syntax beträchtliche Flexibilität und Eleganz, die sie allerdings dem faktischen Verzicht auf eine Oberflächensyntax verdankt. Dementsprechend bleiben oberflächensyntaktische Erwägungen im folgenden außer acht.

3 erstmals entwickelt in VAN FRAASSEN 1969; erweitert und ausführlich dargestellt in FINE 1975

4 Beide Typen überschneiden sich offensichtlich stark. Statt zwischen zwei Klassen von Kennzeichnungen zu unterscheiden, sollte man vielleicht von zwei verschiedenen Verwendungen eines einheitlichen Typs definiter NPs reden - wobei die Semantik der CN-Phrase in der Regel eine der beiden Verwendungen nahelegt. Diese Frage betrifft aber die im folgenden diskutierte Problematik nur am Rande.

5 Unabhängig davon können die CN-Phrasen in beiden Fällen syntaktisch als normale Prädikate behandelt werden.

6 z.B. in MONTAGUE 1974

7 was bekanntermaßen unmittelbar ersichtlich wird, wenn man (7) negiert

8 genauer als zwei Präsuppositionen (existence and uniqueness presupposition)

9 Vertretbar wäre die Präsupposition in dieser Formulierung für Kennzeichnungen vom ersten Typ wie <u>der König von Frankreich</u> (und eventuell <u>der König</u>). Die gängigen Interpretationen definiter Deskriptionen werden ja in der Regel anhand plausibler Beispiele des ersten Typs eingeführt.

10 Ausführlicher dargestellt ist sie in CRESSWELL 1973, S.179f.

11 Ihn so zu definieren, ist kein Problem in der λ-kategorialen Syntax. Was die Definition für eine Auswirkung auf die oberflächensyntaktische Behandlung des bestimmten Artikels hat, ist - s.o. Anm. 2 - kein Problem der λ-kategorialen Syntax.

12 Da Kontext alles umgreift, was der Fall ist, schließt die Situationskenntnis des Kommunikationsteilnehmers sein gesamtes Vorwissen ein. Aus demselben Grunde ist - eine endliche Speicherkapazität des menschlichen Gehirns vorausgesetzt - der vollständige Äußerungskontext streng genommen nie gegeben.

13 Nachgeprüft werden kann das natürlich nur anhand der - hier nicht darstellbaren - Formalisierung. Zur formalen Seite sei nur angemerkt, daß die Menge aller Kontextfragmente etwa die Funktion des specification space in FINE 1975 hat.

14 Wie eindeutig die Schätzung sein muß, ist eine außersemantische Frage. Ihre Beantwortung hängt v.a. von der (geschätzten) Wichtigkeit der vermittelten Information für den Adressaten ab. Natürlich wird schon der Sprecher, je nach der Bedeutung, die er seiner Äußerung zumißt, durch mehr oder weniger ausführliche Spezifikationen hinreichende Eindeutigkeit herzustellen versuchen.

LITERATUR

CRESSWELL, M.J. (1973): Logics and languages. London: Methuen

FINE, Kit (1975): "Vagueness, truth and logic". Synthese 30: 265-300

VAN FRAASSEN, B.C. (1969): "Presuppositions, supervaluations and free logic". LAMBERT, K. (ed.): The logical way of doing things. New Haven: 67-91

HAWKINS, John A. (1977): "The pragmatics of definiteness. Part I". Linguistische Berichte 47: 1-27

KARTTUNEN, Lauri (1974): "Presuppositions and linguistic context". Theoretical Linguistics 1: 181-194

LEWIS, David (1972): "General Semantics". DAVIDSON, D./ HARMAN, G. (eds.): Semantics of natural languages. Dordrecht: Reidel: 169-218

MONTAGUE, Richard (1974): "The proper treatment of quantification in ordinary English".MONTAGUE, R.: Formal philosophy. Selected papers. New Haven/ London: Yale Univ. Press: 247-270

PINKAL, Manfred (1977): Kontext und Bedeutung. Ein probabilistisch erweiterter pragmatischer Beschreibungsansatz. Tübingen: TBL Narr

ON PARAPHRASE RELATIONS

Rüdiger Schreyer

1. It is generally taken for granted that speakers of a language can say the same thing in different words and that they can recognize whether two different utterances mean the same thing or not. This paraphrasing ability plays an important part both in linguistic theory and in linguistic practice, notably in psycholinguistics (SLOBIN 1971) and in semantics.

Most linguists seem to agree that the paraphrasing ability is an important explicandum of linguistic theory (e.g. FODOR 1961; KATZ 1961:37; KATZ/FODOR 1961:175; HIŻ 1968; BIERWISCH 1970:167; HUTCHINS 1971:201; LEECH 1974:84-86). Furthermore a good deal of semantic arguments are explicitely based on the observation that two sentences S_1 and S_2 are paraphrases of each other. According to Leech (1974:84) these observations are taken to be self-evident. He counts statements about paraphrase relations (P-relations)[1] among the "basic statements" which the semanticist has to explain "by constructing theories and descriptive rules and categories from which they can be deduced".

A different view is adopted by Chafe (1971:25), who avers that paraphrase is "incomplete and contextually limited" and who therefore doubts whether it makes sense to develop a formalism from which P-relations should be deducible.

2. Chafe's concern is with the assertion that there are sentences which are paraphrases of each other. If this assertion could be proved to be false or unimportant, attempts to understand and explain paraphrase within the framework of a semantic theory would be either a wild goose chase or not worth the effort. Let us therefore have a closer look at Chafe's argument.

With an eye on TG he suggests that "recent linguistic theory has suffered considerably from a most unsubtle view of meaning identity" which he dubs the "X is really Y" syndrome (1970:86 f.). This reproach led Leech (1974:84) to accuse Chafe of regarding other people's intuitions on meaning identity as less reliable than his own. But surely this is not the point: Chafe merely has a different, extremely narrow conception of paraphrase. Although he does not deny the possibility of complete meaning identity he believes that for a great many so-called paraphrases "surface structures do not, in fact, have identical meanings but only very similar meanings" (1970:87).

Thus Chafe basically claims that P-relations are unimportant because they seldom occur. Note, however, that P-relations do not come labelled as such. Whether we call two sentences P-related depends entirely on our conception of P-relatedness. For Chafe, paraphrases must have identical meanings, in the sense of one semantic structure having a multiple phonetic representation (1970:58). This conception is criticised by Taber (1970:88) as extremely stringent:

> ... it would seem for many purposes at least equally useful to emphasize the great degree of similarity between such terms or expressions, as when they contain all the same semantic

components at the conceptual level but differ in
affective or stylistic impact.

Chafe would probably not object to this, for he only denies the frequent occurence of t r u e paraphrase, not however of meaning similarity (CHAFE 1970:87).

Chafe's argument and particularly his use of the epithet t r u e reminds one very much of the synonymy discussion in which some scholars came to the conclusion that t r u e synonymy does not exist because their interpretations of 'synonymy' turned out not to have a counterpart in language. I have argued elsewhere (SCHREYER 1976:3) that this attitude is unreasonable, and that rather than consider one interpretation or other t h e definition of synonymy one ought to try and find a definition for the linguistic phenomenon which is intuitively felt to exist. The same goes mutatis mutandis for the definition of paraphrase.

Smaby (1974:109) suggests that "we do not agree so well on the use of the term paraphrase or mean the same as one might think " and the dispute about the existence or non-existence of paraphrase seems to support this belief. If the term is to be used in linguistics at all it is the task of the linguist to propose a definition of 'paraphrase'. The purpose of this definition is what Naess (1966:37-73) calls "precization". The method is to propose and examine reasonable interpretations of the term 'paraphrase' without, however, forgetting that in ordinary language the term is used to refer to something which is felt to have linguistic reality.

3. Before embarking on a discussion of various attempts at and difficulties of interpreting 'paraphrase' I will introduce some preliminary, informal distinctions which will give a rough indication of how I wish to use the term 'paraphrase', and then explain how this conception of paraphrase may be related to such linguistically important concepts as ambiguity, synonymy, and definition.

Let us agree, for the moment, on the following working definition:

Sentences are P-related if their meanings are very similar.

This definition has at least one merit: It is not open to Chafe's objections. Moreover, the conception of meaning similarity is indispensable to semantics, for semantics could not even get off the ground without it. If all meanings were seen as radically different, semantics would be confined to the one statement: "Meaning A is different from meaning B, C, D ...".

With Katz (1961:39-40) we may distinguish three types of P-relations:

G-P-relations

Sentences are G-P-related if they are P-related by virtue of their grammar alone, e.g.

(1a) Yesterday I went to see my oculist.
(1b) I went to see my oculist yesterday.

E-P-relations

Sentences are E-P-related if they are P-related by virtue of synonymous[2] expressions occuring in identical places of an otherwise identical grammatical structure, e.g.

(2a) Yesterday I went to see my oculist.
(2b) Yesterday I went to see my eye-specialist.

G-E-P-relations

Sentences are G-E-P-related if they are P-related by virtue both of their grammatical structure and the occurence of synonymous expressions in appropriate places of that grammatical structure, e.g.

(3a) Yesterday I went to see my oculist.
(3b) I went to see my eye-specialist yesterday.

G-P-relations have been amply discussed in GTG. G-E-P-relations are derivable from G-P-relations and E-P-relations and therefore of no theoretical interest. The emphasis in the following sections will be on E-P-relations, although a large part of the general discussion will be relevant to G-P-relations as well.

4. In order to understand the importance of this wider concept of P-relatedness it is instructive to examine the way assertions about P-relatedness are used in linguistic arguments. Linguists, like everybody else, "argue from certain kinds of observations which they take to be self-evident" (LEECH 1974:84). These basic statements are in need of detailed investigation. But it seems to me that assertions about P-relatedness are used more frequently as basic statements than is generally realized. I will demonstrate this by adducing a few examples.

4.1. Ambiguity

Many linguists appear to accept as a basic datum the ambiguity of certain sentences. Chafe, for instance, asserts that ambiguities are very common (1971:9-10). He claims that "ambiguities are objectively observable", by which he seems to mean that different conceptual structures can be recognized by introspection (1970:77; 1971:22). Katz/Fodor (1963:187 ff.) share the opinion that ambiguity is an important feature of language. They give recognition to this by making ambiguity one of the foremost explicanda of a semantic theory. To them it is a necessary, although not a sufficient condition for the semantic ambiguity of a sentence that it contains an ambiguous lexical item.

The question is how do we know that a lexical item or an expression or a sentence are semantically ambiguous? Katz/Fodor (1963) remain silent on that point, whereas Chafe has faith in introspection. On the other hand there is Hiż (1968:212) who stresses that people often fail to realize when a sentence is ambiguous and that it is better not to rely on mere feeling.

Semantic practice proves that the statement "S is ambiguous" is rarely used as basic. It is generally supported by assertions about P-relatedness. Even Chafe adopts this method, as can be gathered from the following example:

> ... given the surface structure Peter is standing up, I think every speaker of English who can understand the issue will agree that its underlying meaning may be either that Peter is in a standing position or that he is

in the process of assuming a standing position, and
that these two are discretely different meanings. (CHAFE 1970:77)

The form of the argument is as follows: A sentence S has two paraphrases, namely S_1 Peter is in a standing position and S_2 Peter is in the process of assuming a standing position. S_1 and S_2 are not paraphrases of each other. Therefore S is ambiguous (HIŻ 1968:212; MORAVCSIK 1972:448). Hiż (1986:218) even suspects "that every pair of paraphrases is used in explaining some ambiguity, so that each of them is actually needed for the clarification of our utterances". Since assertions about ambiguity can thus always be reduced to assertions about paraphrase Leech (1974:89) suggests that ambiguity is "probably best excluded from the categories of basic statement".

4.2. Synonymy

So far no convincing analysis of synonymy has been presented and I am not going to present one either. I will merely point out the close connection between synonymy and E-P-relations.

Probably the most popular definition of synonymy in linguistics runs something like this:

> Expressions are synonymous if they can be interchanged without altering the meaning of all (some of) the sentences in which they occur.

Both the stronger and the weaker version of this definition employ the interchangeability criterion and thus do not accept synonymy as a basic datum. All they really do is describe a method of rephrasing statements about the synonymy of expressions as statements about the synonymy of sentences. Thus they merely reverse the definition of E-P-related sentences given above. They claim:

> If two sentences S_1 and S_2 have the same grammatical structure with the exception that an expression E_1 occurs in S_1 where an expression E_2 occurs in S_2 and if S_1 and S_2 are P-related, E_1 and E_2 are synonymous.

Whereas in E-P-related cases everything depends on the unanalysed notion of P-related expressions (KATZ 1961:39), in this type of definition of synonymy everything depends on the unanalysed notion of P-related sentences. I cannot go into the problems created by this definition of synonymy (see PALMER 1976:63; SCHREYER 1976). But it should be obvious that the definition of E-P-relatedness and the above definition of synonymy are only two sides of the same coin. The definition of E-P-relatedness posits as fundamental the synonymy relation, the definition of synonymy posits as fundamental P-relatedness.

4.3. Definition

One may distinguish between different types of definitions depending on purpose and method (ROBINSON 1954). The purpose of lexical definitions such as can be found in monolingual dictionaries is, to quote Locke (1961 II: 19), "to make another understand by words what idea the term defined stands for". Definitions effected by purely verbal means consist of a definiendum (the term defined), a definiens (the defining expression) and the "meaning relational phrase" (HIORTH 1959:304), which is, however, usually implicit. Hiorth (1959:305) avers that it would be easy, for English alone, to obtain a list of 50 meaning relational phrases. An examination of dictionary definitions shows that the implicit meaning relational phrases may be very dissimilar. Nevertheless something like "means the same as" will probably fit the majority of definitions.

This brings us back to the problem of sameness – or, if you like – close similarity of meaning, this time between a definiendum, which is usually one word, and a definiens, which is usually a longer expression. The definiens may be seen as an extended synonym of the definiendum (PSHENICHNAYA 1967:17). There is no reason why the relation between definiens and the definiendum should not be considered a synonymy relation and thus all that has been said about the connection between synonymy and paraphrase will also apply in the case of verbal definition. Leech (1974:205) seems to share this opinion.

Formalized semantic descriptions in terms of semantic features or meaning postulates must also, in a sense, mean the same as the analysed words, expressions, or sentences. One may, therefore, ask whether they, too, could not ultimately be reduced to the notion of paraphrase (see COOPER 1974). But a discussion of this intricate problem would lead into deep waters and therefore I will have to be satisfied with just drawing attention to it.

4.4. Interim Results

I have tried to justify the claim that the description and explanation of paraphrase is one of the tasks of semantics. The meaning of 'paraphrase' as a pre-theoretical term is vague. Its linguistic relevance depends on the interpretation given to it. Chafe is probably right in maintaining that paraphrases in his interpretation are extremely rare. This interpretation therefore has little interest for linguistics. From this point of view a broader definition of 'P-related' as 'having closely similar meanings' seems preferable to Chafe's very narrow definition. The broader definition may be, and often is, linked with the notions of ambiguity, synonymy, and definition. Assertions about these are rephraseable as assertions about P-relations, although it must be emphasized that the exact formulation of these assertions is very much in need of further investigation. But if it is agreed that ambiguity, synonymy, and definition are important semantic concepts the importance of paraphrase can hardly be denied.

5. A possible objection to the broader definition of 'paraphrase' may be that it is not clear what is being claimed by asserting that speakers can recognize whether sentences have closely similar meanings, and that 'close similarity of meaning' is in need of further precization. This is undoubtedly correct, but there are good reasons for initially retaining this definition.

Ziff (1964:214) writes that in linguistic inquiries there comes a point where one "has to leave the prim path of rosy speculation and muck about with the data". In the case of paraphrase, however, the central problem seems to be that uncontroversial data do not exist. The question is how they can be obtained.

Which data are relevant to a given field of inquiry necessarily depends on certain shared, pre-theoretical assumptions about this field. In the case of paraphrase the highest common denominator seems to be that 'paraphrase' refers to a relation of close similarity obtaining between the meanings of phonetically distinct sentences. The claim is that speakers can recognize this relation. Now, meanings cannot be observed, nor can the knowledge of meaning similarities. The semanticist cannot find out anything about them unless he asks a speaker or a group of speakers.

It is well-known that speakers' intuitions about their use of language are fallible. But although they must never be taken on trust "there is no reason to doubt that there is some degree of correspondence between the way we speak and the way we think we speak" (LYONS 1977:27). In the case of paraphrase speakers' intuitions is all we have to go by.[3] An investigator will have to ask questions or give instructions. These will have to be couched in ordinary language, or he will run the risk of not being understood. Thus, in order to obtain data, an investigator must employ the vague, pre-theoretical terms of ordinary language, because, paradoxical as it may sound, they are the most likely to be understood by his informants.

6. A glance at tests concerned with P-relations shows that although analysts are aware of the vagueness of the notion 'sameness of meaning' (SLOBIN 1971:87) their tests set paraphrasing tasks in terms whose meanings are sometimes even less certain. Smith (1974:11), for instance, asked young children to "say the same thing in different words". Gleitman and Gleitman (1970) instructed their adult subjects to give for each item of a list of compound nouns a phrase that "meant about the same thing" (108:233), or to select from a collection of phrases the one that "meant most nearly the same thing" as the compound they would be given (132:234).

These tests usually achieve their purpose. Hiż (1968:211) remarks that although we do not get reliable answers to the question as to what the meaning of a given phrase is we can obtain reliable reactions to questions concerning the identity or difference of the meaning of two phrases. Speakers apparently have no qualms in discounting the nicer distinctions of meaning which led Chafe to deny the existence of paraphrase.

7. Despite all the theoretical and practical problems connected with eliciting techniques the procedure outlined above seems to me preferable to one that begins with more sophisticated definitions whose translation into ordinary language is uncertain and whose validity cannot be tested. Smith and the Gleitmans, for instance, have far more elaborate conceptions of paraphrase than may be gleaned from their test instructions.

> Sentences are taken to be paraphrases if they have the same underlying grammatical relations and the same cognitive content or truth value. (SMITH 1974:11)
>
> ... paraphrase, a sentence with the same meaning as another sentence, a sentence assigned the same deep structure. (GLEITMAN & GLEITMAN 1970:3)

Whether their test instructions are really translations of these definitions is a matter of dispute. But it can be shown that definitions of 'paraphrase' are often unsuitable for obtaining a reliable set of paraphrases. It will be instructive to examine from this point of view some definitions of 'paraphrase' found in the literature.

> ... one semantic structure having a multiple phonetic interpretation ... (CHAFE 1970:58)
>
> ... different surface structures reflecting the same semantic structure ... (CHAFE 1971:11)

... true synonymy, based on a largely identical set of semantic features ... (TABER 1974: 90)

... sentences are paraphrases of each other if they can be derived from the same semantic representation. (HUTCHINS 1971:222)

Zwei Sätze sind nur dann synonym, wenn sie dieselben Wahrheitsbedingungen haben und dieselben Inferenzen erlauben. (BOURSTIN 1976:247)

X is synonymous with Y if X has the same truth value as Y. (LEECH 1974:86)

Definitions containing the terms 'deep structure' and 'surface structure' are unsuitable because they presuppose the existence of paraphrases: The assumption that P-related sentences exist and the attempt to account systematically for their relation was one of the prime motives for introducing the distinction between deep and surface structure. Furthermore it must be remembered that expressions such as 'deep structure', 'surface structure', 'underlying grammatical relation', 'semantic representation' and 'semantic feature' only make sense within the framework of specific linguistic theories. Useful though they may be in their proper context, they are useless for obtaining data as they cannot be used in testing. They cannot be understood by informants and being theory-specific they cannot be translated into ordinary language without recourse to the whole theory.

For all these reasons I believe that 'close similarity of meaning' is a useful definition to start with, useful for the purpose of eliciting reliable data, i.e. a set of sentences between which this similarity relation is asserted to obtain by a significant number of informants. This is, of course, only the beginning of an investigation of paraphrase, but at least we have reached a point where we have some data to "muck about with". On the basis of these data new hypotheses can be formed or tested, i.e. they can serve as an inspiration or as a test of suggested precizations.

8. To conclude with I will discuss one precization that has been suggested in some of the foregoing definitions and elsewhere, viz. equivalence of truth values. Is T-equivalence a reasonable interpretation and at the same time a precization of 'paraphrase' or – more precisely – of 'close similarity of meaning'?

Chafe (1971:11:25) denies that 'having the same truth value' is an admissible interpretation of 'paraphrase'. He argues against the carrying over into linguistics of ideas from philosophy. With reference to the sentences

10.a Bertrand Russell kicked the bucket.

10.b Bertrand Russell died.

he says:

> The idea that sentences have the same meaning if they have the same truth value is a good example. If, for example, 10.a is true whenever 10.b is true and vice versa, that has been seen as sufficient grounds for saying that 10.a and 10.b are synonymous; that they are complete paraphrases; that, in short, they have the same meaning. This critierion seems to have been accepted by linguists who concern themselves with these matters, but so far as I know it has never explicitly been justified.

He concludes that "there is no reason whatever to believe ... that the truth value criterion is adequate to establish the sameness and difference of messages" (1971:11).

Leech (1974:85), on the other hand, adopts the opposite position when he maintains that basic statements (including statements of P-relatedness) are easily translatable into terms of truth and falsehood and therefore lend themselves to tests of validity.

Again the disagreement is about the admissible interpretations of 'paraphrase' ("synonymy" of sentences in Leech's terminology). There are three statements involved:

1) S_1 is a paraphrase of S_2.

2) S_1 has the same meaning as S_2.

3) S_1 has the same truth value as S_2.

To Chafe 2) is an admissible interpretation of 1) while 3) is not. To Leech 3) is an admissible interpre pretation of 1), not however 2) in the narrow sense of Chafe, for Leech (1974:87) makes the proviso that "the truth conditions hold by virtue of the conceptual meaning alone". But, as Palmer (1976: 62) points out, the notion of conceptual meaning is by no means clear. Leech's (1974:26) interpretataion of it as "logical, cognitive, or denotative content" does not help much to elucidate the concept.

But whatever is meant by 'conceptual meaning', it is obvious that Leech's interpretation of 'paraphrase' is wider than Chafe's. This is why the T-equivalence criterion proposed by Leech may be adequate for his own interpretation and inadequate for that of Chafe.

Chafe does not explain what he would consider to be a justification of the T-equivalence criterion. But on the assumption that tests will reveal enough unanimity about the notion 'close similarity of meaning' to identify at least a subset of all paraphrases, the following justification might be envisaged: One could demonstrate that the T-equivalence criterion applies to all clear paraphrases and that it does not apply to any clear non-paraphrases. More precisely, if 'close similarity of meaning' is accepted as a useful and reasonable interpretation of 'paraphrase', one can try and demonstrate that T-equivalence is a necessary and/or sufficient condition of this interpretation.

It would have to be shown that

1) All sentences with closely similar meanings are T-equivalent.

2) No T-equivalent sentences have clearly different meanings.

There is, of course, no ultimate proof that condition 1) obtains, because the number of T-equivalent sentences and the number of sentences with closely similar meanings is indefinite. Furthermore, there is the problem of sentences that express questions, commands, or performative speech acts. They do not have a truth value and they could therefore never enter into P-relationships, an assumption which can shown to be counter-intuitive. But even if we believe that these difficulties can be overcome (e.g. by transferring the findings from T-equivalent sentences to those without truth values), there is still the second condition to be met. Note, that in the case of condition 2) it would be enough to find one pair of T-equivalent sentences with different meanings to prove that T-equivalence is not a sufficient condition to establish close similarity of meaning.

This is exactly where the T-equivalence criterion fails:

> ...on the assumption that sentences derived from definitions

of terms in a natural language like English are analytic, the sentences All brothers are male and All tigers are animals are logically equivalent, since both are true in all possible worlds. Thus synonymy is an even more narrow relationship than that of logical equivalence. (MORAVCSIK 1972:448)

The tacit assumption here is that the two sentences have very dissimilar meanings. Although T-equivalence thus seems to be at best a necessary condition for sentences to have closely similar meanings, it is an interesting criterion. It is testable because it is translatable into ordinary language (LEECH 1974:85-6). And it is useful because it serves to restrict the set of possible paraphrases. This has been recognized by, for instance, Bourstin whose definition employs the T-equivalence condition together with an additional one, namely sameness of inference. But this is just another possible precization of 'paraphrase', which must be subjected to a new validation process.

NOTES

1 Leech (1974) does not use the terms 'paraphrase' or 'P-relation' but his term 'sentence synonymy' refers to the same phenomenon. In his paper I shall reserve 'synonymy' for meaning similarity of expressions shorter than sentences.

2 This is in modification of Katz (1961:38) who employs 'P-related' where I use 'synonymous'. Nevertheless his intended meaning seems to be very similar. He writes: "The notion of P-related expressions may be left open. We may give it some content by thinking of synonymous expressions, theory determined equivalences, stipulative equivalences etc.".

3 There are, of course, ways and means of keeping subjective errors or idiosyncrasies to a minimum (see Bendix 1971; Quirk/Svartvik 1966). One of the most obvious is to ask several speakers. This is one reason why an investigator is ill-advised if he relies too heavily on his intuition as a native speaker. Furthermore, in linguistic testing great pains are taken to keep informants as naive as possible in order to avoid distortion of test results by habituation. A linguist is probably the least naive informant imaginable. It is not unlikely that his interest and his preconceptions will influence the sample sentences he produces, distort his judgement of sentences he finds elsewhere, and blind him to phenomena he does not expect. A linguist acting as his own informantexclusively willat best give an accurate description of his own idiolect, but most likely not even that.

BIBLIOGRAPHY

BENDIX, Edward H. (1971): "The data of semantic description". STEINBERG, Danny D./JAKOBOVITS, Leon A. (eds.) (1971): Semantics. An interdisciplinary reader in philosophy, linguistics and psychology. Cambridge: Cambridge University Press, pp. 393-409.

BIERWISCH, Manfred (1970): "Semantics". LYONS, John (ed.) (1970): New horizons in linguistics. Harmondsworth: Penguin Books, pp. 166-184.

BOURSTIN, Pierre (1976): "Zur Paraphrasierbarkeit von Kausativkonstruktionen mit Totmachen/Töten, Totschlagen/Erschlagen." BRAUNMÜLLER, K./KÜRSCHNER,W. (eds.) (1976): Akten des 10. Linguistischen Kolloquiums, Tübingen 1975, vol. 2. Tübingen: Niemeyer: 245-252.

CHAFE, Wallace L. (1970): Meaning and the structure of language. Chicago & London: The University of Chicago Press.

––– (1971): "Directionality and paraphrase". Language 47: 1-26.

COOPER, David E. (1974): "Do SR's paraphrase sentences? " Foundations of Language 11: 571-74.

FODOR, Jerry A. (1961): "Projection and paraphrase in semantics". Analysis 21: 73-77.

GLEITMAN Lila R./GLEITMAN, Henry (1970): Phrase and paraphrase. Some innovative uses of language. New York: W. W. Norton.

HARRIS, Zellig S. (1970): "Two systems of grammar: Report and paraphrase". HARRIS, Zellig S. (1970): Papers in Structural and Transformational Linguistics. I. Dordrecht: Reidel: 612-692.

HIORTH, Finngeir (1959): "Origin and control of meaning hypotheses". Lingua 8: 294-305.

HIŻ, Henry (1968): "The role of paraphrase in grammar". O'BRIEN, Richard J. (comp.) (1968): Georgetown University round table selected papers on linguistics 1961-1965. Washington: Georgetown University Press: 211-218.

KATZ, Jerrold J. (1961):"Areply to 'Projection and paraphrase in linguistics'". Analysis 22: 36-41.

KATZ, Jerrold J./FODOR, Jerry A. (1963): "The structure of a semantic theory". Language 39: 170-210.

LEECH, Geoffrey N. (1974): Semantics. Harmondsworth: Penguin Books.

LOCKE, John (1961): An essay concerning human understanding. Edited with an introduction by John W. Yolton. 2 vols. London & New York: Dent and Dutton. (1st ed. 1690).

LYONS, John (1977): Semantics. I. Cambridge etc.: Cambridge University Press.

MORAVCSIK, J. M. E. (1972): Review of : LEECH, Geoffrey N. (1970): Towards a semantic description of English. Bloomington: Indiana University Press. Language 48: 445-454.

NAESS, Arne (1966): Communication and argument. Elements of applied semantics. London: Allen &Unwin.

NOLAN, Rita (1970): Foundations for an adequate criterion of paraphrase. The Hague: Mouton.

PALMER, Frank R. (1976): Semantics. Cambridge: Cambridge University Press.

QUIRK, Randolph/SVARTVIK, Jan (1966): Investigating linguistic acceptability. The Hague: Mouton.

PSHENICHNAYA, L. E. (1967): "Meaning of terms as represented in the information retrieval system of the Institue of Cybernetics of the Academy of Sciences of the UkrSSr. Construction of a semantic tree of term definitions." Automatic Documentation and Mathematical Linguistics 1: 17-21.

ROBINSON, Richard (1954): Definition. Oxford: Clarendon Press.

SCHREYER, Rüdiger (1976): "Synonyms in context". Reproduced by L.A.U.T. (Linguistic Agency at Trier).

SLOBIN, Dan I. (1971): Psycholinguistics. Glenview & London: Scott, Foresman & Co.

SMABY, Richard M. (1971): Paraphrase grammars. (Formal Linguistics Series, vol. 1). Dordrecht: Reidel.

––– (1974): Review of Nolan (1970). General Linguistics 14: 107-112.

STAAL, J. F. (1967): "Some semantic relations between sentoids". Foundations of Language 3: 66-88.

TABER, Charles R. (1974): Review of CHAFE (1970). Linguistics 123: 81-95.

UNGEHEUER, Gerold (1969): "Paraphrase and syntaktische Tiefenstruktur". Folia Linguistica 3: 178-227.

WALPOLE, H.R. (1941): Semantics, the nature of words and their meaning. New York.

ZIFF, Paul (1965): "About what an adequate grammar couldn't do". Foundations of Language 1: 5-13.

4. AUTOMATISCHE ÜBERSETZUNG

ZUR AUTOMATISCHEN PARAPHRASIERUNG NATÜRLICH-SPRACHLICHER SÄTZE

Christopher Habel / Arno Schmidt / Heinz Schweppe

1. Zur linguistischen Semantik

Eines der Hauptziele der Linguistik ist es, die Beziehung zwischen einer Wortkette und ihrer Bedeutung (falls eine solche existiert) zu klären. In dieser Formulierung des Problems sind mehrere nicht unwesentliche Idealisierungen enthalten, so: i. soll etwa durch die Verwendung von "Wortkette" darauf hingewiesen werden, daß wir es hier mit dem Spezialfall geschriebener - nicht gesprochener - Sprache zu tun haben. Dies hat den Vorteil, daß die komplexen Probleme der Phonologie entfallen; andererseits gehen wesentliche Informationen, wie sie etwa durch die Intonation gegeben sind, verloren; ii. wird dadurch, daß nur die Beziehung zwischen Wortkette und Bedeutung untersucht wird, der situative Kontext einer Äußerung vernachlässigt; d.h. pragmatische und deiktische Einflüsse bleiben unberücksichtigt.

Das Problem, die Bedeutung von "Bedeutung" zu klären, soll und kann hier nicht angegangen werden; wir legen für die weiteren Betrachtungen die Äquivalenz von "Bedeutungszuweisung" und "Übersetzung in eine semantische Repräsentationssprache" zugrunde; hierauf basierend ist also eine Bedeutungszuweisung eine Abbildung t (siehe 1):

(1)

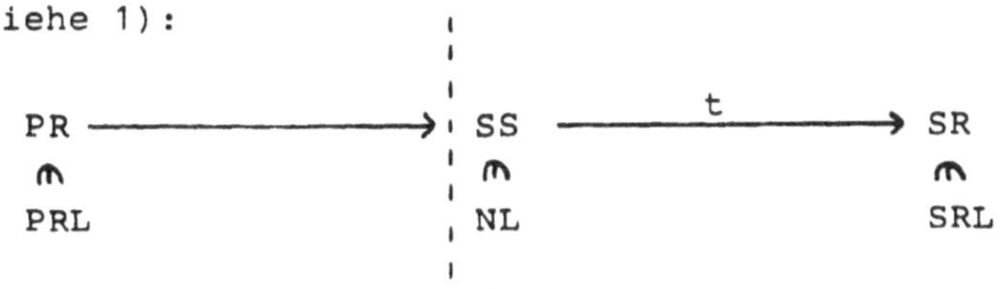

PR = phonologische Repräsentation
PRL = phonologische Repräsentationssprache
SS = Oberflächen-String
NL = natürliche Sprache
SR = semantische Repräsentation
SRL = semantische Repräsentationssprache

Im Hinblick auf die oben erläuterten Idealisierungen betrachten wir das gekennzeichnete Teildiagram. Im allgemeinen ist auch die Umkehrung dieser Abbildungen zu untersuchen, d.h. die Überführung einer semantischen Repräsentation in eine phonologische.

Die in (1) dargestellte Abbildung t findet sich in allen gegenwärtigen Semantiktheorien in einer oder anderer Form, so etwa in der Erweiterten Standard Theorie CHOMSKYs (1967 :420):

(2)

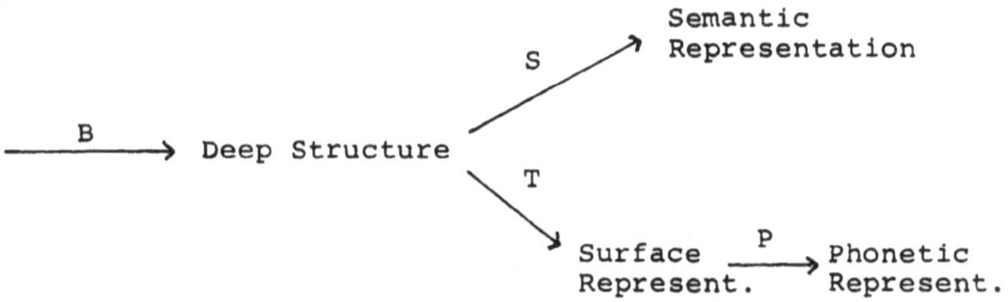

Die Beziehung zwischen (1) und (2) ergibt sich durch die Beziehung der Abbildungen: $t = S \circ T^{-1}$.

Bei der MONTAGUE-Grammatik (1973) wird die Abbildung t durch die simultane Generierung einer (kategorial-) syntaktischen Struktur und eines wohlgeformten Ausdrucks einer formalisierten Sprache der intensionalen Logik, d.h. durch die Korrespondenz von syntaktischen und semantischen Regeln, gegeben.

Die Natürliche Generative Grammatik (NGG) von BARTSCH/VENNEMANN (1972) stellt die Abbildung durch eine Kette von - linguistisch interpretierbaren - Transformationen bereit. Auch Bartsch/Vennemann verwenden eine Sprache der intensionalen Logik als Repräsentationssprache.

Während die NGG und MG über wohldefinierte formale semantische Repräsentationssprachen verfügen, ist dies bei der TG im allgemeinen nicht der Fall, weder in der Ausprägung der interpretativen noch der generativen Semantik. Wie von COOPER/PARSONS jedoch gezeigt wurde, können beide Formen der TG mit einer Sprache der intensionalen Logik als Repräsentationssprache versehen werden, so daß dann TG und MG weitgehend äquivalent sind. Eine starke Kompatibilität zwischen NGG und MG wurde von BARTSCH (1972) nachgewiesen.

Ein wesentlicher Unterschied zwischen NGG auf der einen Seite und MG und TG auf der anderen ist, daß bei TG und MG Produktionsprozesse vorliegen, und zwar bei der MG eine simultane Produktion von syntaktischer und semantischer Struktur (ohne Anspruch auf psychologische Realität), bei der TG (z.B. Chomskys EST) die Generierung einer Tiefenstruktur mit anschließenden interpretativen Komponenten. Die NGG stellt dagegen keine Generierungsprozesse sondern ausschließlich Transformationsprozesse bereit, sie entspricht also weitgehend der schematischen Darstellung des Diagrams 1 (aber mit bijektiven Abbildungen, d.h. Doppelpfeil.).

Das BEAST-System (BErlin Automatic Semantic-oriented Translation)[1] führt die Abbildung t, d.h. den Übersetzungsprozeß von einem natürlich-sprachlichen Satz in einen Ausdruck einer semantischen Repräsentationssprache automatisch, d.h. durch einen Computer, durch. Aus dem bisher Gesagten läßt sich erkennen, daß BEAST in vielen Punkten den Prinzipien der NGG folgt.

Bei einer derartigen Übersetzung ist es im Sinne des Diagramms (1) zu untersuchen, ob t eine Abbildung ist, d.h. ob einem Satz SS genau eine semantische Repräsentation SR zugeordnet wird (dies fällt mit dem Problem lexikalischer und struktureller Ambiguitäten zusammen). Im Falle der EST (Abb. 2) liegen z.B. dann Ambiguitäten vor, wenn T^{-1} keine Abbildung ist, d.h. wenn T keine umkehrbare Abbildung ist, oder anders ausgedrückt, wenn vermittels der syntaktischen Transformationskomponente T keine eindeutige syntaktische Analyse durchgeführt werden kann. Ebensowenig lösen MG und NGG die Probleme der eindeutigen syntaktischen Analyse, bzw. der eindeutigen Strukturzuordnung.

Da es für ein Maschinen-Programm wichtig ist, (Abb. 3) daß eine eindeutige Zuordnung durchgeführt werden kann, da es insbesondere nicht wünschenswert ist, ausschließlich eine Menge von möglichen semantischen Repräsentationen bereitzustellen, wird in BEAST über einen Klärungsdialog die Eindeutigkeit der semantischen Repräsentation erreicht. Die Übersetzung t"(SS) = SR, d.h. die semantische Repräsentation, wird dem Benutzer zur Überprüfung vorgelegt, und zwar in einer der natürlichen Sprache angelehnten Übersetzung $t'(SR_i) = SR'_i$ dieser Repräsentation. Falls diese desambiguierte Darstellung nicht den Intentionen des Benutzers entspricht, d.h. falls das System fehlerhaft oder die Ambiguität anders auflösend übersetzt hat, wird, falls vorhanden, eine alternative Darstellung angeboten. Somit steuert der Benutzer die Übersetzung zu einer eindeutigen semantischen Repräsentation, d.h. er sichert die Abbildungseigenschaft von t.

(3)

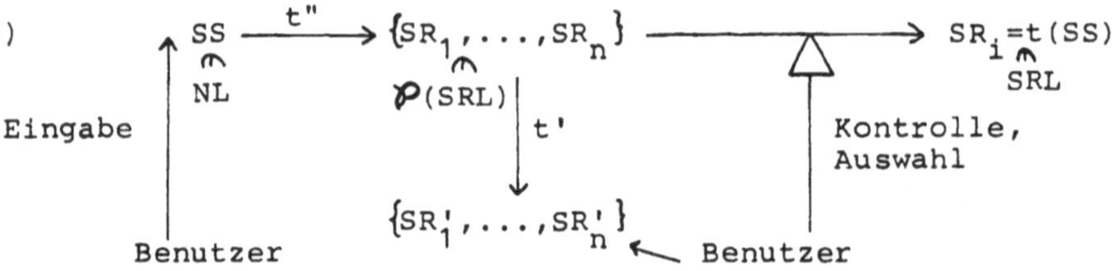

Das "Verstehen" eines Satzes ist mit der Erstellung einer semantischen Repräsentation nicht abgeschlossen, vielmehr muß diese noch interpretiert werden. Den meisten sprachverstehenden Systemen der Künstlichen Intelligenz (KI) - so auch BEAST - und auch den Semantiktheorien der Linguistik, soweit sie über eine Sprache der intensionalen Logik als Repräsentationssprache verfügen (s.o.), liegt folgendes gemeinsames Schema der "possible world semantics" zugrunde:

(4)

$$\mathrm{NL} \xrightarrow{\;t\;} \mathrm{SRL} \xrightarrow{\;i\;} \mathcal{P}(\mathrm{PW})$$

i = Interpretationsabbildung
PW = Menge der möglichen Welten

Die hier vorliegende Abbildung i leistet - intuitiv beschrieben - folgendes: Ist $SR \in SRL$, so ist i(SR) die Menge aller möglichen Welten, in denen SR wahr (bzw. erfüllt) ist, wobei "SR ist wahr" sich auf die Denotate bezüglich der entsprechenden Welt bezieht. Im Sinne der modelltheoretischen Interpretation (HINTIKKA, 1969) wird i durch die Interpretationsabbildung (aus dem Individuenbereich in die mögliche Welt) und die anschließende Auswertung (d.h. Zuweisung eines Wahrheitswertes, meist aufbauend auf dem Fregeschen Prinzip) "induziert".

Im Falle einer einzigen (realen) Welt berechnet sich i durch die Auswertungsvorschrift, d.h. durch die Überprüfung der Wahrheit; im Falle eines Datenbanksystems durch die Anfrage an bzw. den Eintrag in die Datenbank.

Um einen Satz SS $\in$ NL "verstehen" zu können, müssen im Sinne von (4) die Abbildungen t und i durchgeführt werden. (Auf die Schwierigkeiten, überhaupt die Abbildungseigenschaft von t zu erzwingen, haben wir schon hingewiesen). Darüberhinaus genügt es nicht, daß t bzw. i intuitiv oder auch formal berechenbar sind, sondern es besteht vielmehr ein legitimes wissenschaftliches Interesse, Algorithmen für die Berechnung, d.h. Ausführung der Abbildungen zu erhalten. Die Forderung nach einer expliziten Angabe derartiger Algorithmen ist die Forderung nach einer experimentellen Überprüfung linguistischer Theorien. Solange eine explizite Angabe von t und i aussteht, kann von einer Überprüfung oder gar Bestätigung der entsprechenden Theorie nicht gesprochen werden. (WILKS, 1974).

Zur Algorithmisierung von t: Die Programmierung, bzw. die explizite Angabe in einem Regelverzeichnis, stellt innerhalb von BEAST (s.u.) die Algorithmisierung der Abbildung t sicher (unter der Voraussetzung, daß der Algorithmus terminiert). Im Gegensatz zu den Transformationsregeln in linguistisch-orientierten Grammatiken, sind Anwendungsbedingungen und Ergebnisse formal definiert, daher ist ihre Wirkung berechenbar, die Regeln sind anhand einer großen Anzahl von Beispielen nachprüfbar (nicht beweisbar ist die Adäquatheit). Der Definitionsbereich der Abbildung t ist im Fall der Computer-Linguistik - vorerst - auf einfachere Satztypen beschränkt; darüberhinaus fehlen weitgehend die Möglichkeiten, pragmatische bzw. deiktische Aspekte der Sprache zu berücksichtigen[2].

Zur Algorithmisierung von i: Für die linguistisch-philosophisch-logische Semantik stellt die Algorithmisierung von i aufgrund der Komplexität von $\mathcal{P}$(PW), der Potenzmenge der Menge der möglichen Welten, ein ungelöstes Problem dar; weder die Menge der möglichen Welten wird explizit bestimmt, noch wird eine formale Definition der Interpretationsabbildung (im modelltheoretischen Sinn, aus der sich die Interpretation i berechnet) durchgeführt werden. Die reale Welt der sprachverstehenden Systeme stellt natürlich eine starke Einschränkung im Vergleich zur Menge der möglichen Welten dar, zum einen handelt es sich stets um einen beschränkten Welt ausschnitt, zum anderen fehlen - meist - alternative Welten.

2. Das BEAST-System

Der Übersetzungsprozeß (d.h. die Abbildung t) wird innerhalb des BEAST-Systems in (partielle) Teilabbildungen zerlegt; woraus sich folgender System-Aufbau ergibt:[3]

(5)

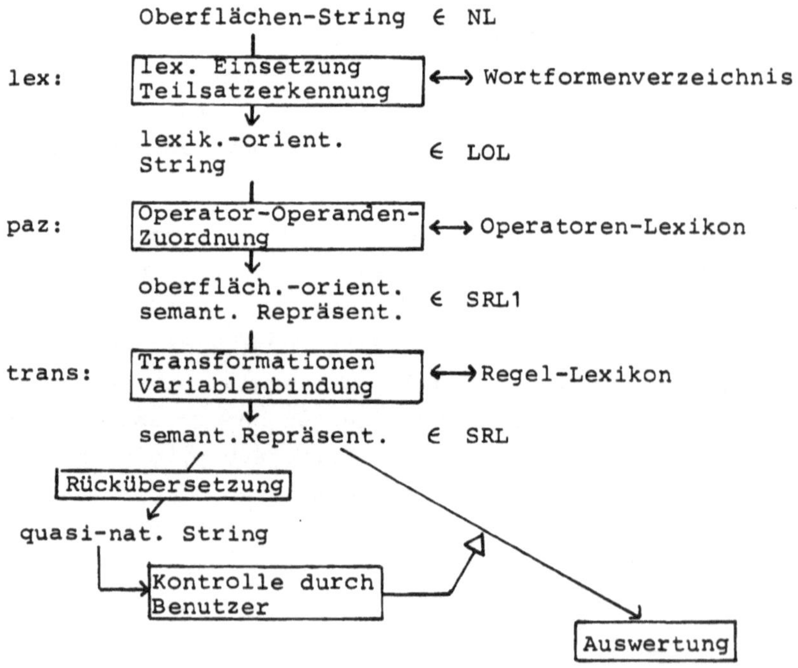

Zur Kommunikation mit dem System verwendet der Benutzer die natürliche Sprache (NL); diese ist nicht formalisiert, implizite Einschränkungen ergeben sich durch den bearbeitbaren Sprachumfang.

Grundlegend für die System-Konzeption ist das P r i n z i p m i n i m a l e r I n f o r m a t i o n. Basierend auf der Tatsache, daß natürliche Sprache redundant ist, d.h. gewisse

Informationen, die für die Bedeutung eines Satzes relevant sind, mehrmals im Satz verschlüsselt sind, gehen wir davon aus, daß eine Übersetzung von der Verwendung von Basisinformationen ausgehen sollte, z.B. die Flektionsformen nur dann in Betracht ziehen sollte, wenn die in ihnen enthaltene Information nicht an anderer Stelle erschlossen werden kann[4].

Auf diesem Prinzip beruhend wird der Analyse-Prozeß der Eingabe unter frühzeitiger Einbeziehung der Semantik, soweit wie möglich oberflächenorientiert, durchgeführt, d.h. ohne Verwendung eines Parsers. Die Abbildung lex ordnet zum einen mit Hilfe eines Wortformenverzeichnisses jedem Wort der Oberfläche eine Kodierung (Basis-Lemma) zu, die dieses semantisch und syntaktisch repräsentiert, zum anderen wird eine Teilsatzerkennung vorgenommen (bisher nur Relativsätze), die den Teilsatz semantisch kennzeichnet[5]. Das Ergebnis von lex ist ein Ausdruck der lexikon-orientierten Sprache (LOL := lex(LOL)).

Das wesentliche Problem der Bedeutungszuweisung (BARTSCH/VENNEMANN) ist, die in der natürlichen Sprache gegebene Operator-Operanden-Beziehung in eine semantisch äquivalente Prädikat-Argument-Beziehung der logischen Sprache zu überführen. Die Abbildung paz leistet unter Verwendung eines Operator-Operanden-Lexikons diese Zuweisung. Dabei sind der Kodierung des natürlichsprachlichen Operators entsprechend seiner Bedeutung verschiedene Prädikate zugeordnet; die Zuweisung geschieht unter Berücksichtigung einer Korrespondenz zwischen der Kombination der LOL-Operanden und den Argumenttyp-Rastern der Prädikate[6]. Das Ergebnis von paz ist ein Ausdruck der oberflächen-orientierten semantischen Repräsentationssprache (SRL1 := paz(LOL)).

Im Gegensatz zu LOL und SRL1 ist die semantische Repräsentationssprache SRL eine formalisierte Sprache (gegenwärtig: SPLL = "standard predicate logic language", von H. ENGEL als Datenbank-Kommunikationssprache definiert)[7]. Die Aufgabe der Abbildung trans besteht darin, die lineare Struktur der SRL1-Ausdrücke in die hierarchische Struktur der Prädikatenlogik zu überführen. Dies geschieht durch die Anwendung von Transformationen, die in einem Regellexikon zusammengefaßt sind und - wie in der Linguistik üblich - aus Strukturbeschreibung und Strukturveränderung bestehen, im Gegensatz zur Linguistik jedoch auf Strings, nicht auf Bäumen, operieren.

Der Sprachumfang des Systems, d.h. die echte Übersetzungskapazität, ist nur durch die Sätze gegeben, die aus NL bis nach SRL überführt werden; die Übersetzung t ist somit auf solche Fälle beschränkt, in denen t(ss)∈SRL gilt, es ergibt sich:

(6) $$t := \text{trans} \circ \text{paz} \circ \text{lex} \,\big|\, (\text{trans} \circ \text{paz} \circ \text{lex})^{-1} \, (\text{SRL})$$

SRL-Ausdrücke werden durch einen Rück-Übersetzer in eine quasi-natürliche Sprache paraphrasiert, um dem Benutzer die Möglichkeit zu geben, zu überprüfen, ob das System seinen Intentionen gemäß übersetzt hat. Ist dies der Fall, so kann er die Auswertung (d.h. Anfrage bzw. Eingabe) veranlassen, anderenfalls eine alternative Übersetzung auf ihre Richtigkeit (bezüglich seiner Intention) überprüfen.

3. Beispiele

NL : Wohin reist Hans ?

LOL :	Wohin	Q003	Kennung für Fragepartikel "wohin"
	reist	§FAHR	Basis-Lemma für: "fahren", "reisen"
	Hans	KT11	Konstante des Objekttyps OT11
	?		

SRL1 :	Wohin	Q002 OT14 Q002 OT13 Q002 OT12	Wohin wurde übersetzt zu: "In welches Land", "In welche Stadt", "Zu welchem Kongreß"
	reist	§REISENLAND(OT11,OT14) §REISENORT(OT11,OT13) §TEILNEHM(OT11,OT12)	Dies sind die SRL-Prädik, die die Bedeutung von §FAHR repräsentieren.
	Hans	KT11	
	?		

In SRL1 sind drei Alternativen angelegt, die durch Argumenttyp-Raster unterschieden werden; im folgenden wird der Transformationsprozeß für die erste Alternative durchgeführt:

```
     Q002 OT14 §REISENLAND(OT11,OT14) KT11 'Hans'
Regel 12:
     KXX(OT14) = §REISENLAND(OT11,OT14) KT11 'Hans'
Regel 18:
     KXX(OT14) = KXX(OT11) .eq. 'Hans' .and.§REISENLAND(OT11,
                                                        OT14)
```

Durch "KXX" werden die Sortenfunktionen einheitlich behandelt; nach dem Transformationsprozeß erfolgt die Namens-Substitution:

```
LAND(OT14) = PERSON(OT11) .eq. 'Hans' .and.
                  REISENLAND(OT11,OT14)
```

Abschließend erfolgt die Variablenbindung:

```
LAND(OT14) = .ex. OT11 (PERSON(OT11) .eq. 'Hans'
             .and. REISENLAND(OT11,OT14) )
```

Quasi-natürlich-sprachliche Paraphrase:

Gib für die Länder OT14, für die der prädikatenlogische Ausdruck wahr ist, die Namen der Länder an.

Prädikatenlogischer Ausdruck:

Es existiert eine Person OT11, für die gilt:
    der Name der Person OT11 ist 'Hans'
 und
    OT11 reist nach OT14

ANMERKUNGEN

1 Das System entstand innerhalb des Projektes "Automatische Erstellung semantischer Netze" am Institut für Angewandte Informatik der Technischen Universität Berlin.

2 Auch innerhalb der Linguistik existieren kaum befriedigende - zumindestens im Hinblick auf die Formalisierung - Lösungen pragmatischer Problemstellungen.

3 Eine ausführliche Beschreibung findet sich in den Arbeiten der Autoren (s.Literaturverzeichnis): (1977a) berücksichtigt mehr die Aspekte der Linguistik, (1977b) die der Informatik.

4 So ist z.B. im Satz "Ich spreche" die Information über den Agens zweimal enthalten, durch "ich" und in der Endung "-e" des Verbs. Die Vernachlässigung der Flektion ermöglicht die Verarbeitung von "deviant utterances"; "du sprechen" wird genauso interpretiert wie "du sprichst".

5 Die Kapazität des Systems kann hier durch Lemmatisierungsalgorithmen und weitere Teilsatz-erkennende Prozeduren erweitert werden.

6 In einer weiteren Ausbau-Stufe des Systems wird durch die Verwendung von Tiefenkasus (in Verbindung mit Objekttyp-Hierarchien) und Wortstellung der Sprachumfang vergrößert werden.

7 Eine Erweiterung zu einer Sprache mit (Modal-) Operatoren ist geplant; hieraus werden sich auch Veränderungen in der Transformationskomponente ergeben.

LITERATUR

BARTSCH, R. (1972): "Relative adjectives and comparison in a Montague grammar". BARTSCH, R./VENNEMANN, T. (1972).

BARTSCH, R./VENNEMANN, T. (1972): Semantic structures. Frankfurt/M.: Athenäum.

CHOMSKY, N. (1967): "The formal nature of language". LENNEBERG, E. (1967): Biological foundations of language. New York.

COOPER, R./PARSONS, T. (1976): "Montague grammar, generative semantics, and interpretative semantics". PARTEE, B. (ed.) (1976): Montague grammar. New York: Academic Press.

ENGEL, H. (1976): "A data query language". Semantic Network Project Report 6/76. Berlin.

HABEL, Ch./SCHMIDT, A./SCHWEPPE, H. (1977a): "On automatic paraphrasing of natural language expressions". Semantic Network Project Report 3/77. Berlin.

--- (1977b): "BEAST - Ein Übersetzungsystem zur natürlichsprachlichen Kommunikation mit Datenbasen". Semantic Network Project Report 4/77. Berlin.

HINTIKKA, J. (1969): "Semantics for propositional attitudes". DAVIS/HOCKNEY (eds.) (1969): Studies in philosophical logic. Dordrecht: Reidel.

MONTAGUE, R. (1973): "The proper treatment of quantification in ordinary English". HINTIKKA, J./MORAVCSIK, J./SUPPES, P. (eds.) (1973): Approaches to natural language. Dordrecht: Reidel.

WILKS, Y. (1974): "One small head - Models and theories in linguistics". Foundations of Language 11.

ZUR STRUKTUR DES ÜBERSETZUNGSSYSTEMS S A L A T[1]

Robert Maier

1. Automatische Übersetzung und Sprachbeschreibungstheorien

Der um das Jahr 1950 aufblühende Optimismus hinsichtlich der schnellen Realisierung eines vollautomatischen Übersetzungssystems hoher Qualität ("fully automatic high-quality translation") ist längst einer realistischeren, eher skeptischen und bisweilen resignierenden Einstellung gewichen. Insbesondere im Zusammenhang mit dem Problem, unter den möglichen Lesarten eines zu übersetzenden Textstückes die jeweils zutreffende herauszufinden, bestätigte sich die alte Erkenntnis, daß zu einer guten Übersetzung zumindest in schwierigeren Fällen das inhaltliche Erfassen des Textes erforderlich ist und hierzu außer einer kompetenten Beherrschung von Ausgangs- und Zielsprache auch ein hinreichendes Allgemeinwissen und spezielle Kenntnisse des betreffenden Sachgebietes Voraussetzung sind. Für die Versuche zur Konstruktion eines automatischen Übersetzungsverfahrens bedeutet dies u.a., daß in einem solchen System Möglichkeiten vorzusehen sind, Abhängigkeiten der Bedeutung von Kontexten und Verwendungssituationen zu erfassen, Schlußfolgerungen zur Auflösung von Ambiguitäten zu ziehen und hierbei auch von nicht dem Kontext zu entnehmenden Informationen ("Weltwissen") Gebrauch zu machen, also Sinnzusammenhänge kalkülmäßig zu erfassen und geeignete nicht-kontextuelle Informationen in einer Datenbasis bereitzustellen. Eine Sprachbeschreibungstheorie, in der diese weit über die Beschreibungsmittel traditioneller Grammatiken hinausgehenden Aspekte integrierend dargestellt werden können, existiert bisher nicht. Ihr Fehlen ist das entscheidende Hindernis auf dem Wege zu einer einwandfreien vollautomatischen Übersetzung.

In dieser Situation wird die vorwiegend praxisorientierte Forschung, der es auf die möglichst schnelle Erstellung wenigstens für bestimmte Zwecke verwertbarer Verfahren ankommt, nicht ein vollautomatisches Übersetzungsverfahren hoher Qualität sondern vielmehr eine möglichst günstige Form der Arbeitsteilung zwischen Mensch und Maschine anstreben. Für so reduzierte Ansprüche an die Leistungsfähigkeit lassen sich durchaus automatische Übersetzungsverfahren entwickeln (und sind in einsatzfähiger Form entwickelt worden), die etwa dort, wo es auf die schnelle Verarbeitung großer Textmengen ankommt, mit menschlichen Übersetzern konkurrieren können (vgl. BRUDERER 1977).

Die vorwiegend theorieorientierte Forschung wird sich dagegen hinsichtlich einer schnellen praktischen Einsatzfähigkeit der von ihr entwickelten Verfahren mit bescheideneren Ansprüchen begnügen müssen. Sie wird ihre Arbeit vielmehr verstehen als einen Beitrag zur Entwicklung leistungsfähiger Theorien der effektiven Sprachbeschreibung und des Sprachvergleiches durch die Erprobung und Überprüfung solcher Theorien im Rahmen eines Übersetzungssystems. Dies zwingt zu einer Darstellung sprachlicher Fakten in einer der formalen Bearbeitung zugänglichen Form und stellt daher hohe Anforderungen an die Klarheit der Problemformulierungen und die Präzision der Begriffsbildungen.[2] Untersuchungen dieser Art bleiben auch dann sinnvoll, wenn man eine vollautomatische Übersetzung hoher Qualität nicht für realisierbar hält. Denn in ihrem Rahmen können sowohl die eventuellen Grenzen der automatischen Übersetzung und ihre linguistischen Gründe exakter erkannt und beschrieben werden als auch die Punkte konkreter erfaßt werden, in denen eine Theorie der Sprachbeschreibung (noch) nicht genügend leistungsfähig ist. Bei der vorwiegend theorieorientierten Entwicklung des Übersetzungssystems S A L A T wird versucht, diesen Gesichtspunkten Rechnung zu tragen.

2. Die Konzeption des Übersetzungssystems S A L A T

Das Verfahren S A L A T soll sich nicht auf die speziellen Probleme und Möglichkeiten der Übersetzung zwischen zwei vorgegebenen Sprachen konzentrieren. Vielmehr geht es darum, die generellen Anforderungen an ein einwandfreies vollautomatisches Übersetzungsverfahren zu klären. S A L A T soll daher unabhängig von einzelnen Sprachen entwickelt werden. Sprachdaten (wie Grammatikregeln) und die Bestandteile der eigentlichen Übersetzungsprozedur (wie Algorithmen zur Anwendung von Grammatikregeln eines bestimmten Typs) werden daher strikt getrennt. Regeln sind also nicht als diskontinuierliche Elemente implizit in Prozeduren enthalten. Dies hat auch Konsequenzen für die Wahl der Programmiersprache.[3] Zur Kontrolle des Verfahrens werden Sprachdaten mehrerer Sprachen verwendet.[4]

Die Übersetzung erfolgt unter Verwendung eines allgemeinen semantischen Repräsentationssystems als Interlingua. Dies wäre bei Beschränkung auf ein Sprachenpaar nicht unbedingt erforderlich. Das semantische Repräsentationssystem dient auch der Darstellung von "Weltwissen" in der Datenbasis. Es muß daher die Darstellung der semantischen Repräsentanten für die Ausdrücke verschiedener natürlicher Sprachen gestatten und eine an den Schlußregeln der Logik orientierte kalkülmäßige Erfassung von Sinnzusammenhängen ermöglichen. Ausgangspunkt für seine Konstruktion ist eine spezielle ε-λ-kontextfreie Sprache (siehe unten), deren von Urs Egli und Arnim von Stechow entwickelte Konzeption eine Kombination von Ideen von Suppes, Cresswell und Montague ist (vgl. EGLI 1975, STECHOW 1974). In Analogie zur Verallgemeinerung der kategorialen Sprachen zu den λ-kategorialen Sprachen durch Cresswell wird eine kontextfreie Syntax um Regeln zur Einführung der Abstraktoren λ und ε für Funktionen und Klassen erweitert. Die so erweiterte Syntax erzeugt Baumstrukturen, die als Darstellung von typentheoretischen Ausdrücken in Baumform aufgefaßt werden können. Sie werden in Anlehnung an die Methode von Suppes, die sich an der modelltheoretischen Interpretation der Typentheorie orientiert, semantisch interpretiert. Für die Computereingabe werden die Strukturen in Klammerschreibweise dargestellt.

Die Überführung ausgangssprachlicher Texte in semantische Tiefenstrukturen (Analyse) sowie die Überführung der Tiefenstrukturen in Sätze der Zielsprache (Synthese) erfolgt mit Hilfe von Transformationsregeln. Hierbei wird eine Modifizierung des Ansatzes zur Darstellung von Transformationsregeln in EGLI 1975 verwendet, die flexibler ist als die aus der Literatur bekannten Transformationsregeldefinitionen. Die Anwendung der Regeln wird durch eine Prozedur zur Erzeugung aller mit Hilfe einer Menge von Transformationsregeln ableitbaren Strukturen geleistet, die auch für Deduktionstransformationen in anderen Teilen von S A L A T verwendet wird. Bei der Analyse werden zunächst für alle in Frage kommenden

Lesarten des Ausgangssatzes Tiefenstrukturen erzeugt, aus denen dann in einer sich anschließenden eigenen Desambiguierungsphase die zutreffende ausgewählt wird. Ein direkter Übergang von Zwischenstrukturen der Analysephase zu Zielsprachenstrukturen - etwa um ein zum Ausgangssatz strukturähnliches Übersetzungsresultat zu erhalten - ist vorerst nicht vorgesehen. Aus Effizienzgründen wird der erste Teil der Analyse-Transformationen in Form eines zweistufigen aus Wortsyntax und Satzsyntax aufgebauten kontextfreien Parsers dargestellt, der zunächst eine Oberflächenstruktur für den Ausgangssatz erzeugt (vgl. BROCKHAUS 1975). Die Entscheidbarkeit wird für das gesamte Analysesystem nicht durch die Eigenschaften der Regelanwendungsprozedur gewährleistet, sondern durch eine intrinsische Regelordnung der Analyse-Transformationsregeln bewirkt, die auf der Verwendung von disjunkten Teilalphabeten auf den jeweiligen "Stufen" des Analyseprozesses beruht. Diese Regelordnung kann zur Effizienzsteigerung (teilweise) explizit gemacht werden. Analyse-Transformationsregeln und Synthese-Transformationsregeln einer Sprache sind nicht identisch oder Umkehrungen voneinander. Bisher wird die starke Analogie zwischen beiden Regelsystemen nur heuristisch ausgewertet. Es scheint jedoch nicht ausgeschlossen, diesen Zusammenhang explizit zu beschreiben und bei der Regelerstellung systematisch zu verwenden. Hierzu sind jedoch zunächst weitere Fortschritte bei der Untersuchung der Möglichkeit nötig, Transformationsgrammatiken als Analysegrammatiken zu verwenden.

Das Ergebnis der Synthese ist eine Struktur der zweistufigen kontextfreien Syntax des Analyseteils; deren Endkette ist das Übersetzungsresultat. Die Synthese erfolgt in zwei formal nicht unterschiedenen, durch die Form des Regelsystems bewirkten Phasen: Zunächst werden bottom-up die anfangs nur bei den Endknoten vorhandenen morphologisch-syntaktischen Informationen auf die höheren Knoten übertragen und schrittweise ergänzt, dann werden top-down die nun vervollständigten syntaktischen Informationen auf sämtliche Knoten übertragen und schließlich die Lexeme an den Endknoten durch flektierte Formen ersetzt. Hierdurch ist es möglich, syntaktische Merkmale erst dann einzuführen, wenn die Informationen für ihre Spezifizierung vorliegen. Auf diese Weise werden Sackgassen vermieden. Da nicht zu jedem Semantem der Interlingua in jeder Sprache ein korrespondierendes Lexem vorhanden ist, wird zwischen Desambiguierung und Synthese ein Transferschritt eingefügt, in dem die Analyseresultate mit Hilfe eines Semantemlexikons in bedeutungsgleiche Tiefenstrukturen transformiert werden, deren Semanteme auf die Zielsprache orientiert sind. So wird sichergestellt, daß zu Beginn der Synthese die Ersetzung der Semanteme durch Zielsprachenlexeme in jedem Fall möglich ist. Für die Entscheidung zugunsten einer bestimmten von mehreren möglichen Umformungen im Rahmen des Transferschrittes werden gegebenenfalls Informationen aus der Datenbasis benötigt; so kann etwa der Fall eintreten, daß aufgrund der aus der Datenbasis entnommenen Eigenschaften des zugehörigen Objektes entschieden wird, ob beim Übersetzen vom Französischen ins Deutsche ein "mettre-Semantem" in ein "stellen-Semantem" oder ein "legen-Semantem" zu überführen ist.

3. Ablaufdiagramm des Übersetzungssystems S A L A T

Die Struktur des Übersetzungssystems S A L A T gibt in einer zum Zwecke der Übersichtlichkeit vereinfachten Form das folgende Ablaufdiagramm wieder. Hierbei werden Daten durch ⟨xxx⟩ und Prozeduren durch [xxx] symbolisiert. Unterbrochene Pfeile geben den Programmablauf wieder, durchgehende Pfeile verbinden Prozeduren mit ihrem Input bzw. Output. Nicht in die Darstellung aufgenommen sind die Umwandlungsprogramme zur Überführung der Daten in die maschineninterne Form und andere Hilfsprogramme wie z.B. Kommandoprozeduren und das Programm zur Überführung des Parser-Outputs in eine für die transformationelle Analyse als Input verwendbare Form. Die Feinstruktur der einzelnen größeren Programmteile und ihr modularer Aufbau aus Teilprogrammen sind nicht erfaßt.

Die Rechtecke in der Mitte des Diagramms stellen die wesentlichen Bestandteile des eigentlichen, von einzelnen Sprachen unabhängigen Übersetzungsverfahrens dar. Transformationelle Analyse und Synthese werden im wesentlichen durch das gleiche Programm zur Anwendung eines Systems von Transformationsregeln auf Baumstrukturen geleistet. Ein auf diesem Programm aufbauendes Deduktionsverfahren ist zentraler Bestandteil der Desambiguierung und des Transfers. Falls bei der Synthese Sackgassen nicht vermieden werden können, ist ein nochmaliger Aufruf des Transferprogrammes mit anschließendem erneuten Syntheseversuch nötig.

Ablaufdiagramm des Übersetzungssystems S A L A T

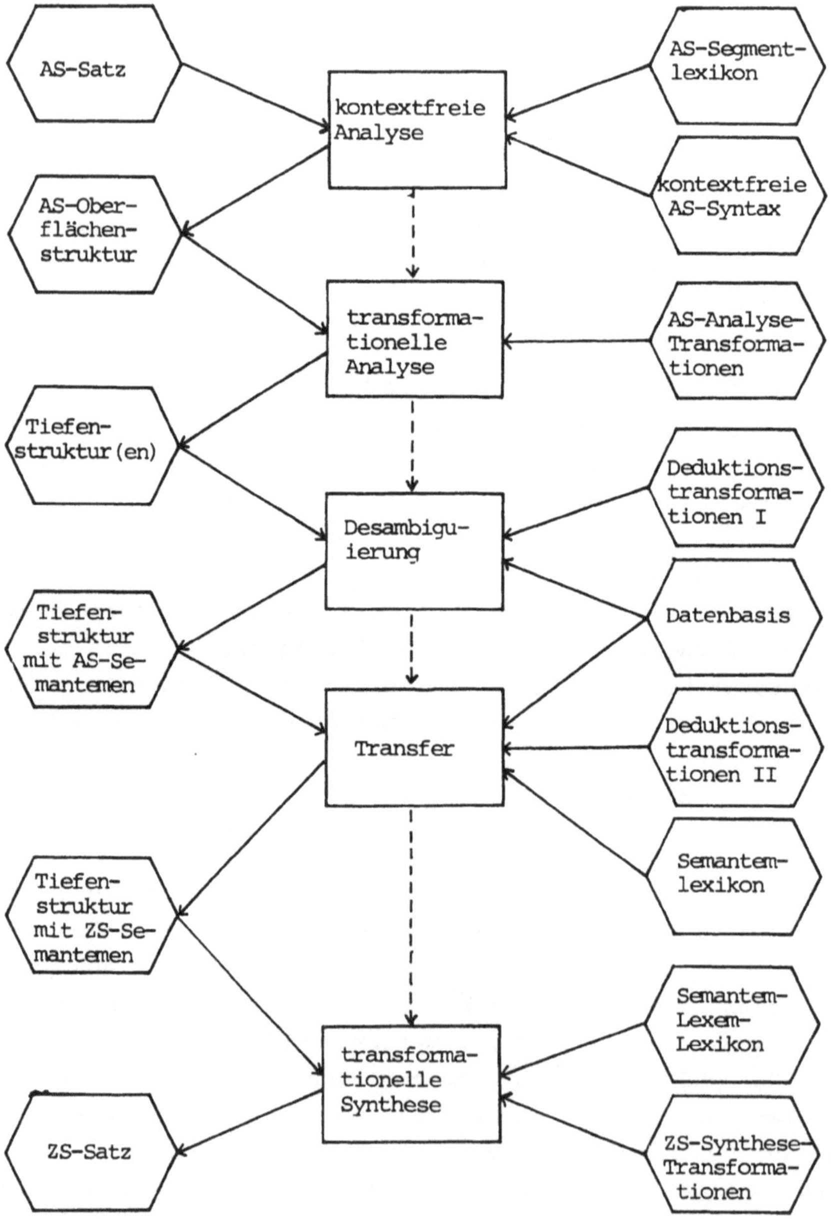

Auf der linken Seite des Diagramms sind die während eines konkreten Laufes benötigten bzw. erzeugten Daten, insbesondere der Ausgangssatz und das Übersetzungsresultat, repräsentiert. Alle diese Daten können in graphischer Baumdarstellung z.B. zu Kontrollzwecken ausgedruckt werden.

Die Daten auf der rechten Seite des Diagramms werden fest auf Datenträgern der Rechenanlage gespeichert. Sie werden für jeden Lauf des Verfahrens benötigt, dabei aber mit Ausnahme eines Teils der Datenbasis nicht verändert. Bis auf die Datenbasis und das Regelsystem für die Deduktionskomponente sind die einzelnen Teile der rechten Seite Sprachbeschreibungsdaten der beteiligten Sprachen

4. Das semantische Repräsentationssystem von S A L A T

Die Kategorien und Regeln der in S A L A T als Interlingua verwendeten ε-λ-kontextfreien Sprache können nicht an den Oberflächenphänomenen einer einzelnen Sprache orientiert sein. Von den technischen Möglichkeiten der ε-λ-kontextfreien Grammatiken sollte im Regelsystem aus methodischen Gründen und z.B. zur Vermeidung zusätzlicher Komplikationen in der Deduktionskomponente nur sparsam und nur dort Gebrauch gemacht werden, wo die darzustellenden sprachlichen Fakten es erfordern. Dies führt zu einer recht einfachen und übersichtlichen Struktur der zentralen Regelgruppe der Basissyntax, mit deren Hilfe u.a. Tiefenstrukturen für Nominalgruppen mit Attributivkonstruktionen und Relativsätzen, komplexe Determinatoren, Verbalgruppen mit verschiedenen Formen der Verbergänzungen einschließlich Nebensätzen, Infinitiven und Adverbien sowie (mit Einschränkungen) Pronominalisierungen darstellbar sind. Die Basissyntax (ohne lexikalische Regeln) in vereinfachter Form, in der die technischen Möglichkeiten nicht enthalten sind, die u.a. zur Darstellung von daß-Sätzen und Tempusphänomen dienen, besteht aus folgenden Regeln:

1. $S \rightarrow MS\ S$
2. $MS \rightarrow KS\ S$
3. $S \rightarrow N\ C1$
4. $N \rightarrow MN\ N$
5. $MN \rightarrow KN\ N$
6. $N \rightarrow Q\ C1$
7. $N \rightarrow X_i$
8. $C1 \rightarrow MC1\ C1$
9. $MC1 \rightarrow KC1\ C1$
10. $C1 \rightarrow X_i\ S$
11. $C1 \rightarrow C2\ N$
12. $C2 \rightarrow MC2\ C2$
13. $MC2 \rightarrow KC2\ C2$
14. $C2 \rightarrow C3\ N$
15. $C3 \rightarrow MC3\ C3$
16. $MC3 \rightarrow KC3\ C3$
17. $MC1 \rightarrow PR1\ N$
18. $PR1 \rightarrow PR2\ N$

Die in der Basissyntax verwendeten Kategorien entsprechen in etwa folgenden Oberflächenkategorien:

S - Satz; N - Nominalgruppen, Eigennamen; Q - Quantoren, Artikel;
C1 - "1-stellige Contentive": Substantive, 1-stellige Verben, durchschnittsbildende Adjektive, Relativsätze, Verbalgruppen;
C2, C3 - "2,3-stellige Contentive": 2,3-stellige Verben;
MS, MN, MC1, MC2, MC3 - 1-stellige Konnektoren bzw. Modifikatoren;
KS, KN, KC1, KC2, KC3 - 2-stellige Konnektoren;
PR1, PR2 - 1,2-stellige Präpositionen.

Regel 10 ist eine abkürzende Schreibweise für die einzige λ-Regel: C1 $\longrightarrow$ λx S. Alle übrigen Regeln außer der Regel 7 zur Variableneinführung sind Zweierregeln, die ein Operator-Operand-Verhältnis ausdrücken: das Denotat der linken Regelseite ist das Ergebnis der Anwendung des Denotats des ersten Teils der rechten Regelseite auf das des zweiten.

Diese Basissyntax kann auch als eine spezielle λ-kategoriale Syntax in Baumschreibweise interpretiert werden. Entsprechend einfach ist die zugehörige modelltheoretische Interpretation: mit Ausnahme der λ-Regel, die wie üblich interpretiert wird (vgl. etwa v. STECHOW 1974), ergeben sich - ausgehend von den den terminalen bzw. präterminalen Knoten zugeordneten Denotaten - die Denotate der höheren Knoten der binären Baumstrukturen jeweils durch Anwendung des Denotats des linken Astes auf das Denotat des rechten Astes. Die Typen der Denotate ergeben sich aus der naheliegenden Festlegung des Typs 0 der Wahrheitswerte für die Kategorie S und des Typs (01) der charakteristischen Funktionen von Klassen von Individuen des Typs 1 für die Kategorie C1. Beispiele für Typenzuordnungen zu den Kategorien sind:

N	- 0(01)	MC1	- (01) (01)
Q	- (0(01)) (01)	KC1	- ((01) (01)) (01)
C2	- (01) (0(01))	PR1	- ((01) (01)) (0(01))

Verschiedene Kategorien gleichen Typs werden (bisher) nicht benötigt. Entsprechend der Vorgehensweise von Montague und Cresswell wird zum Zwecke der einheitlichen Behandlung der Nominalgruppe diese "hochgestuft", so daß sie nicht als Argument, sondern als satzbildender Operator der Verbalgruppe auftritt.

Aus den folgenden Beispielen ist u.a. für Nominalgruppen, Eigennamen, Konnektoren, Relativsätze und Personalpronomen zu sehen, wie in etwa sie in den Basisstrukturen von S A L A T dargestellt werden; Tiefensemanteme sind hierbei zur Verdeutlichung durch flektierte Formen des Deutschen ersetzt worden.

Beispiel 1: Der Hund, der einen Herrn hat, bellt.

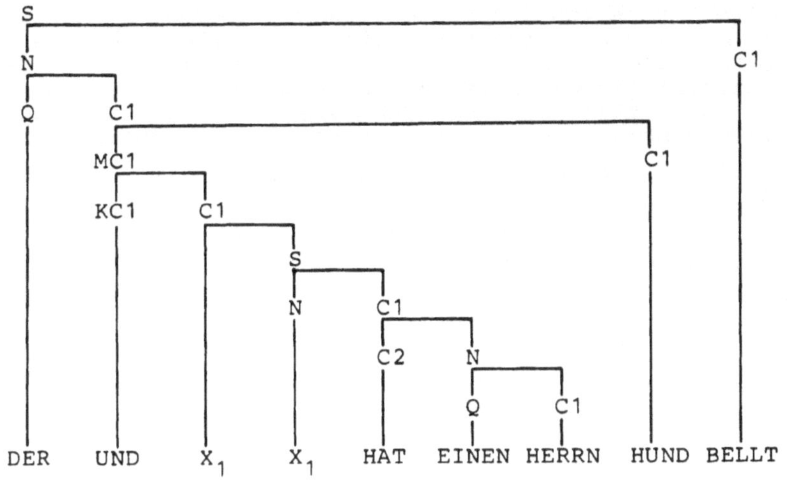

Beispiel 2: Peter trifft Klaus und Ilse und er begrüßt sie.

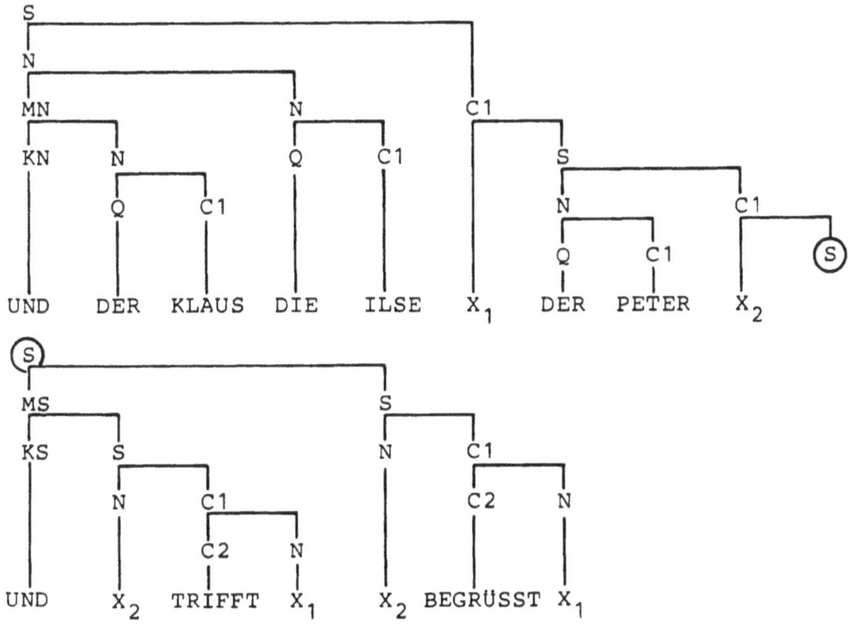

ANMERKUNGEN

1 Das Übersetzungssystem S A L A T (= System for Automatic Analysis and Translation) wird im Teilprojekt A2 ("Automatische Übersetzung") des Sonderforschungsbereichs 99 ("Linguistik") der Deutschen Forschungsgemeinschaft unter Leitung von Prof. Dr. Klaus Brockhaus am Institut für angewandte Sprachwissenschaft der Universität Heidelberg entwickelt. Eine der Hauptaufgaben des Projektes ist die Erprobung der in den in Konstanz durchgeführten übrigen Projekten des Sonderforschungsbereichs entwikkelten Sprachbeschreibungstheorien. Erste Probeläufe des Übersetzungssystems sind erfolgreich durchgeführt worden. Eine ausführliche Darstellung des Übersetzungssystems und seiner Teile wird in BROCKHAUS et al. 1976 gegeben. Sie entspricht allerdings nicht in allen Punkten dem neuesten Stand der Entwicklung.

2 Zu den vorteilhaften Auswirkungen dieses Zwanges vgl. HAUENSCHILD 1978 und HUCKERT 1978.

3 Dies war ein entscheidender Grund, der Programmiersprache PL/I den Vorzug vor LISP zu geben.

4 Zur Zeit werden Sprachdaten für die Sprachen Deutsch, Englisch, Französisch und Russisch verwendet.

LITERATUR

BROCKHAUS, Klaus (1975): "Ein Syntaxformalismus für Flexion und Wortbildung." Linguistische Berichte 40: 45-62

BROCKHAUS, Klaus / HAUENSCHILD, Christa / HÖFLING, Jürgen / HUCKERT, Edgar / MAIER, Robert (1976): Forschungsbericht 1. 11. 1973 - 31. 3. 1976. Sonderforschungsbereich 99 ("Linguistik"). Teilprojekt A2 ("Automatische Übersetzung"). Universität Heidelberg

BRUDERER, Herbert (1977): Handbuch der maschinellen und maschinenunterstützten Sprachübersetzung. München: Dokumentation

CONTE, Maria-Elisabeth / GIACALONE RAMAT, Anna/ RAMAT, Paolo (eds.) (1978): Akten des 12. Linguistischen Kolloquiums. Pavia 1977. Tübingen: Niemeyer

EGLI, Urs (1975): Zur Integration der Semantik in die Grammatik. (masch. Habil.-Schrift Bern 1973). Erschienen als: Ansätze zur ... Kronberg (Taunus): Scriptor

HAUENSCHILD, Christa (1978): "Probleme der automatischen semantischen Analyse des Russischen. Eine geeignete Thema-Rhema-Theorie." CONTE / GIACALONE RAMAT / RAMAT (eds.)

HUCKERT, Edgar (1978): "Automatische Synthese französischer Nominalgruppen." CONTE / GIACALONE RAMAT / RAMAT (eds.)

STECHOW, Arnim von (1974): "ε-λ-kontextfreie Sprachen: Ein Beitrag zu einer natürlichen formalen Semantik." Linguistische Berichte 34: 1-33

PROBLEME DER AUTOMATISCHEN SEMANTISCHEN ANALYSE DES RUSSISCHEN: EINE GEEIGNETE THEMA-RHEMA-THEORIE

Christa Hauenschild

Der Beitrag soll zeigen, wie sich aus einem speziellen linguistischen Problem - nämlich dem der definiten vs. indefiniten Interpretation russischer Nominalgruppen - im Rahmen des automatischen Übersetzungssystems S A L A T (= System for Automatic Language Analysis and Translation, vgl. MAIER 1978) eine ganz konkrete Aufgabenstellung ergibt, die ihrerseits zu einer begründbaren Entscheidung für eine spezielle Version einer Thema-Rhema-Theorie führt.

1. Das Problem

Da das Russische keine Äquivalente zum bestimmten und unbestimmten Artikel in Artikelsprachen (= Sprachen mit obligatorischen Artikeln) hat, ist es bei der Übersetzung in eine solche Sprache erforderlich, die Information über die definite vs. indefinite Interpretation der Nominalgruppen aus dem Kontext zu rekonstruieren, sofern nicht lexikalische Elemente mit einer dem Artikel vergleichbaren Funktion (z.B. Demonstrativpronomina, das Zahlwort "eins" - im folgenden anschaulich, wenn auch nicht ganz zutreffend "Artikelersatz" genannt) die Definitheit vs. Indefinitheit anzeigen. Ich verwende die Begriffe "Definitheit" und "Indefinitheit", ohne sie weiter zu problematisieren, so, daß sie für die Gesamtbedeutung des bestimmten vs. unbestimmten Artikels im Deutschen (und wahrscheinlich auch in einigen anderen Sprachen) stehen, wobei allerdings idiomatische und vorläufig auch generische Verwendungsweisen ausgeklammert bleiben.

Zum Problem der Artikelrekonstruktion, wie ich es im folgenden abgekürzt nennen will, existieren verschiedene, allerdings m. W. nicht sehr zahlreiche, kontrastive Arbeiten (z.B. BRAND 1976, GLADROW 1972, JAKSCHE 1975, KRUŠEL'NICKAJA 1961, Kap. II). Dort werden u.a. folgende Phänomene genannt, die auf die Interpretation von Nominalgruppen bezüglich Definitheit/Indefinitheit Einfluß haben:

Wortstellung und Intonation (oder: die Thema-Rhema-Gliederung der Sätze), Verbaspekt, bestimmte Typen von Attributen, die sog. lexikalischen Mittel, die etwa dem entsprechen, was hier "Artikelersatz" genannt wird. Letztere bestimmen die Interpretation i.a. eindeutig, so daß das eigentliche Problem der Artikelrekonstruktion entfällt. Das Wirken und Zusammenwirken der übrigen Faktoren ist nicht so klar und wird dementsprechend auch nicht so explizit beschrieben.

Hervorzuheben ist der Aufsatz BRAND 1976, weil dort der Versuch einer Algorithmisierung der Artikelrekonstruktion bei der automatischen Übersetzung aus dem Russischen ins Deutsche unternommen wird, allerdings auf einer im wesentlichen syntaktischen Grundlage. Aufgrund einer sehr sorgfältigen Materialanalyse und weitgehenden Ausnutzung der deutschen Idiomatik werden recht gute Ergebnisse erzielt (Fehlerquote bei einem wissenschaftlichen Text: 2-3%!); das Problem kann aber nicht grundsätzlich gelöst werden.

Im Rahmen des Übersetzungssystems S A L A T , das nicht zu möglichst schnell verwertbaren Ergebnissen führen, sondern vielmehr eine grundlegende Untersuchung der anstehenden Fragen ermöglichen soll, ist eine solche Lösung daher nicht akzeptabel. Hier wird die Hypothese aufgestellt, daß das Problem auf der Grundlage einer geeigneten Thema-Rhema-Theorie (genauer: einer speziellen Version einer solchen Theorie) behandelt werden kann, was durch die Tatsache gestützt wird, daß in Arbeiten zur Thema-Rhema-Problematik in Artikelsprachen häufig eine Verbindung zwischen Thema-Rhema-Gliederung und Artikeln hergestellt wird, die allerdings nicht in einer einfachen Korrelation besteht.

2. Die Aufgabenstellung im Rahmen von S A L A T

Wie bereits aus dem Titel des Beitrags hervorgeht, stehen Fragen der A n a l y s e im Mittelpunkt, wobei mit "automatischer semantischer Analyse" die Übersetzung von syntaktischen Oberflächenstrukturen in Ausdrücke des als Interlingua fungierenden semantischen Repräsentationssystems (= Basis) gemeint ist. Fragen der S y n t h e s e (= umgekehrte Übersetzungsrichtung) werden nicht behandelt.

Das Problem der Artikelrekonstruktion tritt in S A L A T schon bei der Analyse - also unabhängig von der jeweiligen Zielsprache - auf, weil hier mit einer logik-orientierten Basis gearbeitet wird, die obligatorisch Artikelrepräsentanten (= Quantoren, die allerdings im Gegensatz zu üblichen prädikatenlogischen Notationen keine satzbildenden, sondern nominalgruppenbildende Operatoren sind) enthält.

Wenn nun eine Thema-Rhema-Theorie als tertium comparationis gewählt wird, sind an diese im Rahmen von S A L A T folgende Anforderungen zu stellen:

a. sie muß die Zusammenhänge der Oberflächenphänomene, die auf die definite vs. indefinite Interpretation russischer Nominalgruppen Einfluß haben, explizit machen, und zwar so explizit, daß
b. die Darstellung in der Basis von S A L A T , deren Ausdrücke modelltheoretisch gedeutet werden können, möglich ist und daß
c. automatisch verarbeitbare Transformationsregeln für den Übergang von syntaktischen Oberflächenstrukturen zu Ausdrücken der Basis formuliert werden können, die aufgrund einer Hierarchie der zusammenwirkenden Faktoren für jeden genügend Information enthaltenden Einzelfall zu einer Entscheidung führen, welche Art von Quantor einer bestimmten Nominalgruppe in der semantischen Repräsentation zugeordnet werden soll.

Wie aus dem Flußdiagramm in MAIER 1978 hervorgeht, schließt sich an die transformationelle Analyse in S A L A T eine Desambiguierungsstufe an, die zwischen mehreren möglichen Lösungen die richtige(n) heraussuchen soll. Zur automatischen semantischen Analyse im oben erläuterten Sinne gehören beide Schritte, so daß eine wichtige Aufgabe bei der Aufstellung der Faktorenhierarchie darin bestehen wird, zu entscheiden, welche Faktoren bereits durch die transformationelle Analyse erfaßt werden können. Das sollten möglichst viele sein, da es sich im Vergleich zur Desambiguierung um den einfacheren Algorithmus handelt, weil weder Deduktionssystem noch Datenbasis-Informationen benutzt werden müssen.

Im folgenden soll als erster Teil der Aufgabenstellung die Entscheidung für eine geeignete Version einer Thema-Rhema-Theorie behandelt und deren Darstellung in der Basis von S A L A T skizziert werden.

3. Eine geeignete Thema-Rhema-Theorie

Der mir bekannte Teil der Diskussion um die Thema-Rhema-Gliederung von Sätzen bzw. Äußerungen (wobei ich hier die terminologischen Differenzierungen nicht berücksichtigen will) kreist besonders im Anschluß an SGALL/HAJIČOVÁ/BENEŠOVÁ 1973 u.a. um die Frage, welche der verschiedenen prinzipiellen Definitionsmöglichkeiten die geeignete ist. Das gilt z.B. für die Aufsätze BARTSCH 1975, DAHL 1975, ESSER 1976, WEISS 1975.

Man kann wohl von drei Grund-Ansätzen ausgehen (vgl. WEISS 1975: 25):

a. die auf Aristoteles zurückgehende Dichotomie zwischen dem, worüber man spricht, (Thema) und dem, was man darüber aussagt, (Rhema);
b. die in der Prager (genauer: Brünner) Schule besonders durch Jan Firbas auf der Grundlage von a. entwickelte Theorie des relativen kommunikativen Gewichtes der einzelnen Konstituenten eines Satzes;
c. die bei Sgall zugrundeliegende Dichotomie zwischen kontextuell gebundenen und nicht kontextuell gebundenen Elementen, die etwa der schon früher (u.a. durch Mathesius) in die Diskussion eingebrachten Unterscheidung zwischen Gegebenem und Neuem entspricht.

Die in SGALL/HAJIČOVÁ/BENEŠOVÁ 1973 vorzufindende Vermischung dieser verschiedenen Definitionen ist in ähnlicher Weise bereits bei Mathesius vorhanden, der nebeneinander a. und c. benutzt (vgl. FIRBAS 1974: 23).

Bei den kritischen Reaktionen auf SGALL/HAJIČOVÁ/BENEŠOVÁ 1973 herrscht Einigkeit darüber, daß diese Vermischung vermieden werden muß, weil sie zu Unklarheiten und Widersprüchen führt. Die Ansichten darüber, welche der Definitionen bzw. welche Kombination aus ihnen zu wählen und wie sie zu präzisieren ist, gehen dagegen wesentlich auseinander, wobei auf verschiedenen Ebenen argumentiert wird. Um zu einer Entscheidung zu kommen, werden u.a. folgende Fragen in den Mittelpunkt der Betrachtung gestellt:

- die Frage der empirischen Überprüfbarkeit, z.B. durch den Fragetest (WEISS 1975, SVOBODA 1974; der zweite Aufsatz ist keine direkte Reaktion auf SGALL/HAJIČOVÁ/BENEŠOVÁ 1973, beschäftigt sich aber mit der hier interessierenden Problematik),
- die Frage der möglichst eindeutigen Definierbarkeit der Grundbegriffe (DAHL 1975, RAIBLE 1971; für letzteren gilt das gleiche wie für den Aufsatz von Svoboda),
- die Frage, ob es sich um eine "semantische" oder eine "pragmatische" Theorie handelt bzw. handeln soll (BARTSCH 1975 auch im Hinblick auf VENNEMANN 1975, ESSER 1976).

Diese Diskussion führt m.E. nicht zu einer begründbaren Entscheidung für eine der angebotenen Definitionen oder für eine Kombination aus ihnen, weil jeweils nur einzelne Aspekte zur Entscheidungsgrundlage gemacht werden. Vielmehr ist von den Anforderungen auszugehen, die sich aus der jeweiligen konkreten Aufgabenstellung ergeben (letztere wird in der Thema-Rhema-Diskussion - und in einem großen Teil der linguistischen Literatur allgemein - allerdings nur selten überhaupt definiert). Das bedeutet natürlich, daß nicht von der geeigneten Theorie schlechthin die Rede sein kann, sondern nur von einer jeweils geeigneten Version.

Für die hier gestellte Aufgabe der Artikelrekonstruktion bei der automatischen semantischen Analyse des Russischen im Rahmen des Übersetzungssystems S A L A T ist die Frage der empirischen Überprüfbarkeit dahingehend zu beantworten, daß die Richtigkeit der Ergebnisse der automatischen Übersetzung in eine Artikelsprache als Kriterium für die Adäquatheit der Theorie benutzt wird. Die Notwendigkeit einer möglichst klaren Definition der Grundbegriffe und der Entscheidung für eine "semantische" Theorie (d.h. für eine Theorie, die sich in dem semantischen Repräsentationssystem darstellen läßt; das ist allerdings ein anderer Gebrauch von "semantisch" als in den angeführten Aufsätzen) ergibt sich zwingend aus den in Abschnitt 2. formulierten Anforderungen.

Daraus folgt, daß eine Version einer Thema-Rhema-Theorie zu wählen ist, die sich nicht schon im ersten Ansatz allzu weit von den zu beschreibenden Phänomenen entfernt. In diesem Sinne bietet sich eine präzisierte Fassung von Definition c. an, die sich an die in

RAIBLE 1971: 212f. für das Französische gewählte anlehnt, jedoch die in SGALL/HAJIČOVÁ/BENEŠOVÁ 1973 für die Definition der kontextuellen Gebundenheit benutzten Grundvorstellungen im Auge behält und dementsprechend die Terminologie wählt:

- die Unterscheidung zwischen kontextuell gebundenen und kontextuell nicht gebundenen Elementen bezieht sich nur auf nominale und pronominale Konstituenten, bei denen nicht generischer Gebrauch vorliegt; auf verbale Konstituenten wird die Dichotomie nicht bezogen;
- kontextuell gebunden sind alle anaphorischen und/oder deiktischen Nominal- oder Pronominalgruppen, d.h. Personalpronomina, Eigennamen sowie Nominalgruppen, die etwas im Situations- oder Textkontext bereits Bekanntes nennen (aufs Russische bezogen: Nominalgruppen mit Possessivpronomina oder Demonstrativpronomina als "Artikelersatz" und solche, die in eine Artikelsprache mit dem bestimmten Artikel übersetzt werden, sofern dort nicht idiomatische Verhältnisse vorliegen; diese Aufzählung ist möglicherweise nicht vollständig);
- nicht kontextuell gebunden sind alle Nominal- und Pronominalgruppen, die weder anaphorisch noch deiktisch sind, also z.B. Indefinitpronomina und Nominalgruppen, die in nicht-idiomatischen Fällen mit dem unbestimmten Artikel übersetzt werden.

Da diese Definition eine wesentliche Einschränkung gegenüber c. beinhaltet, werde ich im folgenden zur Vermeidung von Mißverständnissen nicht mehr die Begriffe "Thema" und "Rhema" benutzen, sondern von der Dichotomie "kontextuell gebunden"/"nicht kontextuell gebunden" sprechen.

Die sich aus dieser Definition ergebende Theorie (die natürlich erst mit der Aufstellung der Faktorenhierarchie vollständig ist, vgl. Abschnitt 2.) hat einerseits eine starke Korrelation zu den behandelten Phänomenen, andererseits ist sie jedoch keine einfache Abbildung: sie erklärt z.B. das gleiche oder ähnliche Funktionieren bestimmter Pronomina und bestimmter Nominalgruppen in gewissen Kontexten (u.a. wenn anaphorische Referenz vorliegt). Die Nähe zu den Phänomenen bewirkt jedoch, daß es in jedem Einzelfall ziemlich einfach sein dürfte, sich darüber zu einigen, wie die in einem Text vorkommenden nominalen und pronominalen Konstituenten hinsichtlich der definierten Dichotomie "kontextuell gebunden"/"nicht kontextuell gebunden" einzuordnen sind. Es handelt sich also wohl um eine hinreichend klare Definition der Grundbegriffe.

Die Theorie läßt sich ohne Schwierigkeiten in der Basis von S A L A T abbilden, wo Pronomina und anaphorische Nominalgruppen als gebundene Variable repräsentiert werden und wo in den Repräsentanten für deiktische und nicht kontextuell gebundene Nominalgruppen Quantoren vorkommen, die den bestimmten vs. unbestimmten Artikel wiedergeben. In der modelltheoretischen Interpretation der Basis-Ausdrücke erscheinen diese Quantoren derzeit als Jota-Operator bzw. als Existenz-Quantor, was vermutlich keine endgültig befriedigende Lösung ist. Die hiermit verbundenen Schwierigkeiten betreffen jedoch nicht die Struktur der Basis-Ausdrücke, sondern lediglich deren modelltheoretische Deutung.

Ein wichtiger Nebeneffekt bei der jetzt gewählten Theorie-Version besteht darin, daß die ganze Problematik der Kontrastierung (oder "second instance", vgl. SGALL/HAJIČOVÁ/BENEŠOVÁ 1973, Kap. 2.4), die noch viel ungeklärter zu sein scheint als die übrigen Teile der angebotenen Thema-Rhema-Theorien, nicht berücksichtigt werden muß, zumindest nicht im ersten Arbeitsschritt (s. u.).

Abschließend bliebe noch die Frage zu untersuchen, ob die nun zugrundegelegten Definitionen zur Behandlung des Problems der Artikelrekonstruktion ausreichen. Es ist ja nicht von vornherein auszuschließen, daß verschiedene Thema-Rhema-Definitionen kombiniert werden müssen in dem Sinne, daß jede für sich präzisiert und dann ihr Zusammenhang (der natürlich nicht in der Identität der Unterscheidungen bestehen darf) explizit gemacht wird. Eine Antwort auf diese Frage kann natürlich solange nur aufgrund von Plausibilitätsüberlegungen gegeben werden, wie die Theorie noch nicht vollständig formuliert ist.

Die meisten der Überlegungen, die i.a. zur Frage der Notwendigkeit einer weitergehenden Theorie angestellt werden, gehen eindeutig über die Erfordernisse der hier gestellten Aufgabe hinaus. Sie sollen jedoch noch auszugsweise betrachtet werden, weil sie für die folgenden Arbeitsschritte im Rahmen des Übersetzungssystems S A L A T relevant werden könnten.

In SGALL/HAJIČOVÁ/BENEŠOVÁ 1973 wird die Behauptung, daß eine Thema-Rhema-Theorie im Sinne von a. bzw. b. in ein semantisches Repräsentationssystem eingebracht werden muß, u.a. mit dem Hinweis auf Skopusprobleme begründet (z.B. auf S. 43). Hierfür ist jedoch eine spezielle Thema-Rhema-Theorie nicht notwendig, da in einem logik-orientierten semantischen Repräsentationssystem ohnehin auf die Wahrung der Skopusverhältnisse zwischen verschiedenen Quantoren Negationsoperatoren und gegebenenfalls Modifikatoren (die in der Basis von S A L A T vorkommen, vgl. MAIER 1978) geachtet werden muß. Die Frage, ob möglicherweise eine Theorie gemäß a. erforderlich ist, um diese Skopusverhältnisse aus den einzelsprachlichen Oberflächen zu analysieren, geht über den Rahmen der Aufgabe der Artikelrekonstruktion hinaus und soll daher vorläufig offenbleiben.

Zu den Beispielen, in denen angeblich Bedeutungen unterschieden werden müssen, ohne daß Skopusunterschiede vorliegen (SGALL/HAJIČOVÁ/BENEŠOVÁ 1973: 43), müßte zunächst klargemacht werden, worin überhaupt der Bedeutungsunterschied bestehen soll (vgl. BARTSCH 1975: 87). Falls das gelingt, dürfte es in einem hinreichend flexiblen semantischen Repräsentationssystem, das nicht auf die Prädikatenlogik der 1. Stufe beschränkt ist, möglich sein, den Unterschied abzubilden, ohne eine vollständige Thema-Rhema-Theorie gemäß a. oder b. zu benötigen.

Zum Schluß sei noch angemerkt, daß es für die automatische - ebenso wie für die menschliche - Übersetzung keineswegs von vornherein als selbstverständlich angesehen werden kann, welche Anforderungen an die Übersetzungsäquivalenz und dementsprechend an die Feinheit des Bedeutungsbegriffes zu stellen sind. M.E. läßt sich ein (zumindest für bestimmte Textsorten) sinnvoller Begriff der Übersetzungsäquivalenz definieren, in dem alle Phänomene, die üblicherweise unter dem Stichwort "Thema-Rhema-Gliederung" (oder entsprechenden) angeführt werden und die über die Wahrung von Skopusverhältnissen und die Beachtung der oben formulierten Dichotomie "kontextuell gebunden"/"nicht kontextuell gebunden" hinausgehen, nicht berücksichtigt werden müssen. Es sollte ein Stufenmodell der

Übersetzungsäquivalenz angesetzt werden (wie es - allerdings mit einer etwas anderen Zielsetzung - in KOMISSAROV 1973 eingeführt wird), in dem sich die jeweils höheren Stufen durch höhere Äquivalenzanforderungen von den tieferen unterscheiden.

Unabhängig davon, wie diese Hypothese beurteilt wird, ist es im Rahmen eines automatischen Übersetzungssystems in jedem Fall sinnvoll, zunächst zu versuchen, mit engeren (und damit i.a. präziseren) Definitionen auszukommen, die allerdings so gewählt sein sollten, daß sie eine spätere Erweiterung nicht unmöglich machen. Das scheint mir bei der hier gewählten Version einer Thema-Rhema-Theorie gewährleistet zu sein; denn sie schließt eine umfassendere Theorie in einem präzisierten Sinne von a. keineswegs aus. Sie schafft vielmehr erst die Grundlage für eine solche Präzisierung, weil die Dichotomie "kontextuell gebunden"/"nicht kontextuell gebunden", wie sie oben definiert worden ist, dabei als ein Faktor unter anderen berücksichtigt werden muß.

LITERATUR

BARTSCH, Renate (1975): "Topik-Fokus-Struktur und kategoriale Syntax". EHRICH, Veronika / FINK, Peter (eds.) (1975): Beiträge zur Grammatik und Pragmatik. Kronberg/Ts.: Scriptor: 85-99

BRAND, Ingeborg (1976): "Die Einsetzung des Artikels bei der automatischen Übersetzung aus dem Russischen ins Deutsche." BRAND, Ingeborg / KLIMONOV, Gerda / LJUDSKANOV, Alexander (1976): Automatische Sprachübersetzung III. Berlin: Akademie: 7-160

DAHL, Östen (1975): "Review of SGALL/HAJIČOVÁ/BENEŠOVÁ (1973)". Journal of Linguistics 11: 347-354

DANEŠ, František (ed.) (1974): Papers on functional sentence perspective. Prag: Academia. The Hague etc.: Mouton

ESSER, Jürgen (1976): "Comment zu einem kontroversen Topic." Linguistische Berichte 46: 71-81

FIRBAS, Jan (1974): "Some aspects of the Czechoslovak approach to problems of functional sentence perspective." DANEŠ (ed.) (1974): 11-37

GLADROW, Wolfgang (1972): "Das Zusammenwirken unterschiedlicher sprachlicher Mittel zum Ausdruck der Determiniertheit/Indeterminiertheit des Substantivs im Russischen." Zeitschrift für Slawistik 17: 647-656

JAKSCHE, Harald (1975): "Zum Ausdruck der Definitheit/Indefinitheit im Deutschen und Russischen." Russisch in kontrastiver Sicht. Symposion Hamburg 1974. Hamburg: Buske: 66-80

KOMISSAROV, V. N. (1973): Slovo o perevode. Moskau: Izd. "Meždunarodnye otnošenija"

KRUŠEL'NICKAJA, K. G. (1961): Očerki po sopostavitel'noj grammatike nemeckogo i russkogo jazykov. Moskau: Izd. literatury na inostrannych jazykach

MAIER, Robert (1978): "Zur Struktur des Übersetzungssystems S A L A T ." CONTE, Maria-Elisabeth / GIACALONE RAMAT, Anna / RAMAT, Paolo (eds.) (1978): Akten des 12. Linguistischen Kolloquiums. Pavia 1977. Tübingen: Niemeyer

RAIBLE, Wolfgang (1971): "'Thema' und 'Rhema' im französischen Satz." Zeitschrift für französische Sprache und Literatur 81: 208-224

SGALL, Petr / HAJIČOVÁ, Eva / BENEŠOVÁ, Eva (1973): Topic, focus and generative semantics. Kronberg/Ts.: Scriptor

SVOBODA, Aleš (1974): "On two communicative dynamisms." DANEŠ (ed.) (1974): 38-42

VENNEMANN, Theo (1975): "Topics, sentence accent, ellipsis: a proposal for their formal treatment." KEENAN, Edward L. (ed.) (1975): Formal semantics of natural language. Cambridge etc.: Cambridge Univ. Press: 313-328

WEISS, Daniel (1975): "Topic und ein seltsamer Comment." Linguistische Berichte 36: 24-36

AUTOMATISCHE SYNTHESE FRANZÖSISCHER NOMINALGRUPPEN

Edgar Huckert

1. Abriß der Problematik

Unter automatischer Synthese wird im folgenden der zweite Teil eines automatischen Übersetzungsvorgangs verstanden. Im ersten Teil, der Analyse, wird für einen Ausgangssatz eine syntaktische Korrektheitsprüfung und die Bereitstellung einer semantisch interpretierbaren Tiefenstruktur geleistet. Die Synthese überführt diese Tiefenstruktur in syntaktisch korrekte und semantisch adäquate Oberflächenstrukturen des Zielsatzes. Die Tiefenstrukturen sind Ausdrücke der semantischen Interlingua des Übersetzungssystems S A L A T[1]. Technisch gesehen sind die Tiefenstrukturen Ausdrücke der Basis einer transformationellen Grammatik. Die Synthese wird vollständig transformationell durchgeführt.

Es geht uns hier um die automatische Synthese von Nominalgruppen, die außer substantivischen Gliedern auch adjektivische Ergänzungen aufweisen. Uns interessiert besonders der Fall der Übersetzung einer deutschen Nominalgruppe mit beliebig vielen Adjektiven. Solche Nominalgruppen sind im Deutschen üblich, etwa:

(1) Eine schöne, sehr junge, schwangere und rothaarige Frau

Im Deutschen können alle Adjektive in nahezu beliebiger Zahl vorgestellt werden. Im Französischen stellt sich die Situation jedoch anders dar; hier muß z.B. zwischen vor- und nachgestellten Adjektiven unterschieden werden:

(2) Une très jeune femme

aber:

(3) +Une enceinte femme

Die Tiefenstruktur der deutschen Nominalphrase kann also nicht geradewegs in französische Oberflächenstrukturen überführt werden. Insbesondere Adjektivhäufungen im Deutschen erfordern komplexe Synthesestrategien. Die traditionellen und modernen französischen Grammatiken drücken sich meist mit vagen Regeln um die Formulierung expliziter Stellungsregeln. So findet man Regeln wie:

> D'une façon générale, on a tendance à placer en tête du groupe l'élément le plus court.[2]

Mit solch vagen Regeln kann man in einer automatischen Synthese kaum etwas anfangen. Es geht also darum, greifbare Kriterien für die korrekte Linearisierung adjektivischer Modifikatoren zu formulieren. Die Regeln sollen zudem in einer Regelsprache formuliert werden, die von einem effektiven Transformationsalgorithmus als Eingabe akzeptiert wird. Im einzelnen stellen sich folgende Probleme:

- Gibt es semantische Kriterien für die Vor- oder Nachstellung von Adjektiven?
- Gibt es syntaktische Kriterien für die Vor- oder Nachstellung?
- Gibt es Interaktion zwischen substantivischen und adjektivischen Modifikatoren bei der Adjektivpositionierung?
- Gibt es Beschränkungen in der Kombinationsfähigkeit bei Adjektivhäufungen?
- Wie werden die gefundenen Regeln konkret formuliert?

Wir vernachlässigen hier lexikalische, pragmatische oder stilistische Probleme bei der Übersetzung. Die Zielsprachen-NP soll lediglich die gleiche Information wie die Ausgangssprachen-NP wiedergeben.

2. Eine semantische Klassifikation der Adjektive

Die Oberflächenkategorie Adjektiv wird in den Tiefenstrukturen von S A L A T durch mehrere Konstruktionen wiedergegeben. Diese Aufspaltung ist semantisch bedingt. Wir unterscheiden:

- durchschnittsbildende Adjektive. Die durch Adj+NP bezeichnete Eigenschaft ist im Durchschnitt der Menge von Dingen, auf die die im Substantiv und die im Adjektiv ausgedrückten Eigenschaften zutreffen, z.B. <u>ein grüner Baum</u>.
- relativierende Adjektive. Sie können wie durchschnittsbildende Adjektive analysiert werden, wenn eine nicht explizit genannte Leerstelle besetzt wird, z.B. <u>eine kleine Maus</u>. Eine Maus kann relativ zur Gattung der Mäuse, zur Gattung der Säugetiere etc. klein sein.
- relationale Adjektive. Sie stehen meist für Gegenstände, nicht für Eigenschaften. Beispiel: <u>une décision ministérielle</u>. In der

Tiefe erscheinen zwei NPs, die durch eine Präposition verbunden sind, in der die Art der Beziehung zwischen den NPs ausgedrückt wird. Sehr viele relationale Adjektive haben auch andere, z.B. durchschnittsbildende Lesarten. Ministériel hat z.B. auch die Lesart "style qui charactérise un ministre".

- privative Adjektive. Dies ist eine wenig homogene und sicher weiter zu analysierende Klasse von Adjektiven. Beispiel: un prétendu assassin. Die semantische Abgrenzung von den anderen Adjektivklassen ist nötig, weil bestimmte Schlüsse nicht zulässig sind, so z.B. der Übergang von "x est un prétendu assassin" auf "x est un assassin". Die Darstellung in der Tiefe ist noch nicht klar. Trotzdem werden diese Adjektive bei den Syntheseregeln berücksichtigt, weil sie sich syntaktisch gesondert verhalten.

Es wird nicht behauptet, daß dies die einzig mögliche und sinnvolle Klassifikation der Adjektive wäre.

3. Stellungsregeln für Adjektive

3.1. Methodisches Vorgehen

In den nachfolgenden Regeln werden semantische und syntaktische Aspekte vermischt, da diese Trennung im Algorithmus primär nicht gegeben ist. Es wird davon ausgegangen, daß semantische Gegebenheiten an den Bäumen ablesbar sind. Die Regeln sind keinesfalls als Definition der zugelassenen Adjektiv-Substantiv-Konstruktionen zu verstehen. Es geht vielmehr darum, aus der Menge derjenigen zugelassenen Konstruktionen diejenigen auszuwählen, die problemlos erzeugt werden können. Die Menge der synthetisierbaren Strukturen ist also eine Teilmenge der analysierbaren Strukturen. Die Synthese liefert als Endprodukt Oberflächenstrukturen, die den Analysebäumen einer kontextfreien Grammatik des Französischen in komplexer Notation entsprechen.

3.2 Vorstellung von Adjektiven

Vorstellbare Adjektive können in einer festen Liste erfaßt werden und werden im Lexikon entsprechend markiert. Nahezu alle vorstellbaren Adjektive können auch nachgestellt werden, so daß sie durch POS(1,2) markiert werden. Vorstellbare Adjektive werden nachgestellt, sobald sie durch bestimmte Adverbien erweitert werden, z.B.

(4) une très jolie fille

(5) une fille extrêmement jolie

Die nachstellungsauslösenden Adverbien wie extrêmement in (5) können in einer festen Liste erfaßt werden. Sie erhalten die Markierung INV(2), während die stellungskonservierenden Adverbien wie très in (4) die Markierung INV(1) erhalten. Die vom Adverb ausgelöste Umstellung (in der Tiefe stehen die Adjektivkorrelate vor dem Substantiv) erfolgt nur dann, wenn die übrige syntaktische Umgebung dies erlaubt. Erweiterungen im nach-adjektivischen Bereich führen immer zur Nachstellung, vgl.

(6) une large fente

(7) une fente large de trois mètres

Diese beiden Regeln können als Syntaktisierungen der in vielen Grammatiken erwähnten rhythmischen Gründe für die Adjektivpositionierung angesehen werden.

Die von der Schulgrammatik her bekannten Regeln für die Adjektiveinstellung bei Bedeutungswechsel, etwa bei ancien, spielen hier keine Rolle, da in der Tiefe eindeutige Semanteme vorliegen, etwa &ALT oder &EHEMALIG. Bei der Ersetzung der Semanteme durch Lexeme wird ancien mit der Stellungsmarkierung POS(1) oder POS(2) erzeugt, je nachdem welches Semantem vorlag.

Die Zahl der vor das Substantiv tretenden Adjektive ist beschränkt. Da keine feste Zahl angebbar ist, setzen wir sie auf zwei fest. Wir lassen also

(8) Une belle jeune fille

zu, nicht aber

(9) +Une belle jeune grande fille

Sind mehr als zwei vorstellbare Adjektive vorhanden, so müssen einige nachgestellt oder durch andere Konstruktionen ersetzt werden. Die beiden vorgestellten Adjektive können konjunktiv verbunden sein, wie in

(10) Une très belle et très jeune fille

In diesem Fall sind beide Adjektive adverbial erweiterbar - unter Berücksichtigung der oben genannten Adverbregel. In der Juxtaposition darf dagegen nur das erste der beiden Adjektive erweitert sein:

(11) Une très belle fille

(12) +Une belle très jeune fille

Die letzte Fügung ist nur mit Komma möglich, also

(13) Une belle, très jeune fille

Die Oberflächenkonstruktion mit Komma verweist auf eine logische Konjunktion in der Tiefe. Alle Juxtapositionen der Oberfläche können auf zwei Konstruktionen in der Tiefe zurückgehen, nämlich Schachtelung und Konjunktion. Die NP (8) kann also auf zwei Tiefenstrukturen zurückgehen, die paraphrasierbar sind als "une fille qui est jolie parmi les grandes filles" und als "une fille qui est jolie et grande".

Außer einigen privativen Adjektiven (prétendu, ancien...) sind in der Vorstellung nur relativierende Adjektive möglich. Hier liegt nach unserer Analyse eine gewisse Ungenauigkeit vor, da eine zu besetzende Leerstelle oft nicht besetzt ist. Dies deckt sich mit der Beobachtung der Grammatiker, daß semantisch gewichtige, präzise Adjektive im allgemeinen nachgestellt werden. Natürlich spielt hier auch die Thema-Rhema-Verteilung im Satz eine Rolle. Diese wird jedoch zur Zeit nicht in der Basis erfaßt.

3.3. Nachstellung von Adjektiven

Einige Regeln zur Nachstellung von Adjektiven ergeben sich aus den Einschränkungen der Vorstellungsregeln. Die meisten Adjektive sind als nur nachstellbar markiert. Bei der Nachstellung müssen vor allem Regeln für die Adjektivkonjunktion, für die Adjektiverweiterung und für die Interaktion zwischen Adjektiven und nominalen Ergänzungen zum Substantiv formuliert werden.

Die Juxtaposition (ohne Komma) von zwei oder mehr nachgestellten Adjektiven ist nicht möglich:

(14) +Une femme enceinte rousse

Dagegen ist Häufung von nachgestellten Adjektiven möglich, wenn sie durch et oder Komma verbunden sind. Wir formulieren folgende Regel: Liegen $n(n>1)$ nachstellbare Adjektive vor, so werden die Adjektive mit den Nummern $n-1$ und n durch et verbunden. Die übrigen Adjektive werden durch Kommata voneinander getrennt. Diese Regel führt zu Oberflächenstrukturen wie:

(15) Une femme jolie, enceinte et rousse

Nachgestellte Adjektive sind im Prinzip beliebig links- und rechtserweiterbar. Dabei müssen jedoch einige Regeln beachtet werden, die das Verhältnis von adjektivischen zu substantivischen Ergänzungen beschreiben. Es gibt lexikalisierte Nominalgruppen, die nicht aufgebrochen werden dürfen:

(16) des pommes de terre farcies
+des pommes farcies de terre

Solche Gruppen erhalten ein Merkmal, das sie als nicht-trennbar markiert. Bei Gruppen, die nicht aus einem einzigen Semantem entstanden sind, hängt die Positionierung des Adjektivs von der Stellung des Adjektivkorrelats in der Tiefe ab. Das Adjektiv tritt hinter die Gruppe S de S, wenn in der Tiefe Determination der ganzen Gruppe vorlag:

(17) une ville de pêcheurs italienne

Diese Regel gilt freilich nur für den Fall, daß keine phonetisch bedingten Mehrdeutigkeiten im Bezug des Adjektivs auftreten, wie in

(18) un village de pêcheurs italien(s)

Falls italien in der Tiefe die gesamte Gruppe S de S determiniert, erzeugen wir die NP un village italien de pêcheurs. Dabei fallen allerdings ebenfalls zwei Deutungen zusammen. In diesem Fall läßt also die Oberflächenstruktur das Nachspielen der tiefenstrukturellen Verhältnisse nicht zu. Die Regel zwingt dazu, die Adjektive im Lexikon besonders zu markieren, deren phonetisches Verhalten in der Flexion wechselt. Bei Fügungen S de Artikel S steht das Adjektiv hinter dem ersten Substantiv, selbst wenn in der Tiefe Determination der ganzen Gruppe vorlag:

(19) la maison grande de mon père

Bei anderen Deutungen sind natürlich auch Stellungen wie in

(20) l'odeur d'un printemps ensoleillé

möglich. Bei Rechtserweiterung des Adjektivs muß dieses generell an das Ende der Gruppe gestellt werden.

Es bleibt die Frage, ob die eindeutigen tiefenstrukturellen Verhältnisse (Schachtelung, Konjunktion) in der Oberfläche nachgespielt werden können. Dies ist sicher nicht der Fall. Schachtelung in der Tiefe, die bei vorstellbaren Adjektiven durch Juxtaposition wiedergegeben werden kann, kann bei nachgestellten Adjektiven nicht ebenso wiedergegeben werden. Alle betrachteten Fälle, in denen mehrere postsubstantivische Adjektive vorlagen, weisen Fügung mit Komma oder Konjunktion auf. Dabei betrachte ich Fälle wie un conseiller juridique français nicht als Ausnahme, da conseiller juridique als lexikalische Einheit aufgefaßt werden kann. Konjunktion in der Tiefe (oder eine ähnliche Konstruktion) kann dann problemlos in der Oberfläche nachgespielt werden, wenn alle an der Konjunktion beteiligten Glieder als Adjektive lexikalisiert werden können. Ist dies nicht der Fall, dann kann die Konjunktion in der Tiefe nicht durch et-Verbindung in der Oberfläche wiedergegeben werden. Schachtelung in der Tiefe kann zum Teil dadurch nachgespielt werden, daß die Glieder mit engerem Skopus als NPs des Typs de S lexikalisiert werden (wenn möglich). Dahinter steckt die in vielen Grammatiken zu findende Hypothese, daß die Gruppe de S enger zum Substantiv gehöre als die übrigen adjektivischen Determinatoren

Die Überlegungen zu Schachtelung und Konjunktion von Adjektiven gelten nur für durchschnittsbildende und relativierende Adjektive. Relationale Adjektive sowie privative Adjektive werden wohl selten in konjunktiver Fügung mit Adjektiven anderen Typs oder gar mit Adjektiven des gleichen Typs auftauchen. Da hier die Analyse als Filter wirkt, brauchen für die Synthese keine besonderen Regeln für diese Adjektive formuliert zu werden.

4. Regelordnung und Regelrestriktionen

Für die algorithmische Umsetzung der aufgestellten Regeln wird davon ausgegangen, daß intrinsische Regelordnung genügt. Die in der Literatur angegebenen Beispiele für die Notwendigkeit extrinsischer Regelordnung oder gar globaler Ableitungsbeschränkungen überzeugen nicht. Sie sind nämlich meist in der Tradition Chomskys geschrieben, d.h. für Oberflächenbeschreibungen, für Tiefenstrukturen wie für Zwischenstrukturen werden die gleichen Kategorien benutzt. Das vorliegende Konzept geht von einer strikten Trennung des Tiefen- und Oberflächenvokabulars aus. Zudem werden komplexe Kategorien, also Hauptkategorien plus Unterkategorien, benutzt, so daß sehr viel Information aus früheren Ableitungsstadien mitgeschleppt werden kann. Der Verzicht auf globale Ableitungsbeschränkungen fällt leicht, da kaum einzusehen ist, wie diese Restriktionen sinnvoll algorithmisiert werden könnten.

Aus den aufgestellten Regeln ist zu ersehen, daß bei der Synthese von einer bestimmten Reihenfolge bei der Regelanwendung ausgegangen werden muß. Wenn z.B. vorstellbare Adjektive nicht mehr vorstellbar sind, sobald sie rechtserweitert werden, dann muß - algorithmisch gesprochen - der Prozeß der Rechtserweiterung vor dem Prozeß der Adjektivpositionierung erfolgen. Auch die adverbiale Erweiterung muß vor der endgültigen Stellungstransformation erfolgen, da die Art der adverbialen Linkserweiterung Konsequenzen für die Stellung hat. Wird einer der Tiefenmodifikatoren als nominale Gruppe z.B. des Typs S de S realisiert, so muß die Positionierung weiterer adjektivischer Modifikatoren Bezug auf die Struktur dieser nominalen Gruppe nehmen. Die Möglichkeit schließlich, tiefenstrukturelle Verhältnisse in der Oberfläche nachzuspielen - z.B. durch explizite Konjunktion - hängt stark von der lexikalischen Realisierung der Tiefenmodifikatoren ab.

Die Diskussion der Regelordnung für das behandelte Problem könnte nun zur Annahme verleiten, hier würde doch eine extrinsische Regelordnung angestrebt. Dies ist nicht der Fall. Das Regelsystem arbeitet nur mit intrinsischer Ordnung, d.h. frühere Transformationen schaffen erst die Anwendungsbedingungen für nachfolgende Transformationen. Auch Zyklen sind durch intrinsische Regelordnung leicht herstellbar. Der Verzicht auf eine extrinsische Regelordnung fiel aus zwei Gründen leicht: einerseits werden die synthetisierbaren Zielstrukturen durch eine vorliegende kontextfreie Analysesyntax des Französischen definiert, die ihrerseits schon ein großes Kategorieninventar umfaßt, andererseits erwies sich bei der Erstellung des Programms eine Stufung des Regelvokabulars sinnvoll zur Erzielung günstiger Rechenzeiten. In einer bestimmten Ableitungsetappe können aufgrund der Kategorien im Ableitungsbaum viele Regeln von vorneherein ausgeschlossen werden, nämlich diejenigen, die Kategorien enthalten, die im aktuellen Ableitungsbaum nicht vorkommen. Durch diese Stufung der Vokabulare während einer Ableitung ist es möglich, die Steuerung des Ableitungsvorgangs dem System selbst zu überlassen.

An einem Beispiel wie

(21) Les deux littératures française et allemande

läßt sich zeigen, wie das grobe Geschütz der globalen Ableitungsbeschränkung einfach vermieden wird. In vielen Ansätzen würde die Blockierung der Numeruskongruenz mit einer globalen Ableitungsbeschränkung erreicht, die man etwa so verbalisieren kann: Wende die Numeruskongruenzregel nicht an, wenn die Koordination mit et durch eine Tilgung einer der beiden identischen Substantiva entstanden ist. Hier wird der gleiche Effekt durch intrinsische Regelordnung erreicht. Geht man davon aus, daß in der Tiefe eine Struktur wie "la littérature française et la littérature allemande" vorliegt, so muß bei der Tilgung eines Vorkommens von "littérature" dafür gesorgt werden, daß die Kategorie von et eine Markierung erhält, die sie von der üblichen Konjunktion unterscheidet. Die Numeruskongruenzregel kann dann später auf dieses Merkmal Bezug nehmen. Diese Lösung verhindert es, daß eine ganze Ableitung verworfen wird. Statt im Nachhinein erlaubt es die vorliegende Lösung, Ableitungen von vorneherein und lokal zu blockieren. Sie hat zudem den Vorteil, daß nicht ganze Ableitungen (von prinzipiell beliebiger Länge) gespeichert werden müssen.

Der technische Ablauf der Synthese ist grob folgender: Zunächst werden für die ausgangssprachlichen Semanteme, die als Endknoten im Analysebaum hängen, französische Lexeme eingesetzt. Dabei wird die Struktur des Analysebaums berücksichtigt, d.h. die Lexikontransformationen sind nicht einfache Ersetzungen des Typs $a \longrightarrow b$. Da ein Semantem oder eine Semantemkombination durch mehrere Lexeme ersetzt werden kann, werden u.U. aus einem Analysebaum mehrere hybride Bäume erzeugt. Während der Ableitung werden die meisten der so entstandenen alternativen Lösungswege eliminiert, weil sich z.B. eine Lexikalisierung eines Semantems als syntaktisch nicht kompatibel mit seiner Umgebung herausstellt. Dann werden die Tiefenkategorien bottom-up durch Zwischenkategorien ersetzt, wobei gleichzeitig die richtige Reihenfolge der Elemente der Zielstruktur erzeugt wird. Dabei können Knoten erzeugt werden, die rein technische Bedeutung haben, z.B. Marken zur Erzielung der richtigen Regelabfolge. Es entstehen so hybride Bäume, d.h. Bäume, in denen gleichzeitig Analyse- und Synthesekategorien vorkommen. Den Abschluß dieser

Zwischenphase bildet ein nicht-hybrider Baum in dem nur noch Kategorien des Zwischenvokabulars vorhanden sind. Auch die technischen Hilfsknoten sind verschwunden. Die alternativen Ableitungsstränge sind bis auf einen einzigen eliminiert worden. Danach werden die Zwischenkategorien top-down durch Kategorien einer kontextfreien Syntax ersetzt, die die Zielsprachenoberflächenstruktur definiert. Umstellungs- oder Tilgungstransformationen finden in dieser top-down Phase nicht mehr statt. Das Ergebnis dieser Phase - der morphologischen Synthese - ist ein vollständiger Analysebaum, dessen Endketten die Morpheme der Ziel-NP sind. Das Synthesekonzept sieht keine Sackgassen vor.

ANMERKUNGEN

1 s. BROCKHAUS et al. 1976 und MAIER 1978.

2 GRAMMAIRE LAROUSSE 1964: 205

LITERATUR

BROCKHAUS, Klaus / HAUENSCHILD, Christa / HÖFLING, Jürgen / HUCKERT, Edgar / MAIER, Robert (1976): Forschungsbericht 1. 11. 1973 - 31. 3. 1976. Sonderforschungsbereich 99 ("Linguistik"). Teilprojekt A2 ("Automatische Übersetzung"). Universität Heidelberg

GRAMMAIRE LAROUSSE (1964): Grammaire Larousse du Français Contemporain. Paris: Larousse 4. Aufl. 1971

MAIER, Robert (1978): "Zur Struktur des Übersetzungssystems S A L A T ." CONTE, Maria-Elisabeth / GIACALONE RAMAT, Anna / RAMAT, Paolo (eds.) (1978): Akten des 12. Linguistischen Kolloquiums. Pavia 1977. Tübingen: Niemeyer

VERZEICHNIS DER AUTOREN

BÄUERLE, Rainer (Universität Konstanz, Fachbereich Sprachwissenschaft, 7750 Konstanz, Postfach 7733) geb. 1949 in Stuttgart. Studium der Germanistik und Anglistik in Tübingen und Konstanz. Wissenschaftlicher Angestellter am Fachbereich Sprachwissenschaft der Universität Konstanz. Hauptarbeitsgebiete: Tempus/ Frage-Antwort. Publikationen: "Tempus, Temporaladverb und die temporale Frage" (Teil 1 und 2, Papiere des SFB 99, Konstanz 1977); "Tempus, Zeitreferenz und temporale Logik. Eine Bibliographie 1940-76" (Linguistische Berichte 49, 1977); "Fugitive propositions again" (erscheint in Analysis); "Tense logics and natural language" (erscheint in Synthese).

CARACCIOLO, Piero (via Tonezza, 14, 0019 Roma; rue Jacob, 13, 75006 Paris) geb. 1950 in Rom. Studium der Philologie und Sprachwissenschaft in Rom und Paris VII. Lecteur d'italien, Université de Lille III. Hauptarbeitsgebiete: Semantik, Sprachphilosophie, Psychoanalyse und Linguistik. Publikationen: "Teorie linguistiche ed epistemologia marxista" (in"Linguistica, semiologia, epistemologia", SLI, Roma, Bulzoni, 1972); "'Questo' e 'quello' in italiano"(1976).

CONTE, Maria-Elisabeth (vgl. das Herausgeberverzeichnis).

DELBECQUE, Nicole (Fakulteit Wijsbegeerte en Letteren, Afdeling Toegepaste Linguïstiek, Blijde-Inkomstraat 21, 3000 Leuven, Belgien) geb. 1952 in Kortrijk. Studium der romanischen Philologie und Hispanistik in Leuven. Aspirante N.F.W.O. (Belgian National Science Foundation). Hauptarbeitsgebiet: Spanische Linguistik. Publikation: "El sujeto y el predicado con los verbos copulativos: un análisis experimental de 'Cancionero' de Miguel de Unamuno." (Cuadernos del Centro de Estudios Hispánicos de la Universidad de Lovaina, 1977).

DEZSŐ, László (Department of Linguistics, Kossuth University, 4010 Debrecen, Hungary) geb. 1927. Studium des Russischen und Englischen in Budapest. Professor für Sprachwissenschaft an der Universität Debrecen. Hauptarbeitsgebiete: Typologie, Fremdsprachendidaktik, Slawistik. Publikationen (im Gebiet der Typologie auf englisch): "Introducing Typology to Contrastive Linguistics" (The Hague, 1978); "Studies in Syntactic Typology and Contrastive Linguistics" (Budapest, 1978); "Typological Studies in Old Serbo-Croatian Syntax" (Budapest, 1978).

FAVA, Elisabetta (via Ortigara,13, 36078 Valdagno (Vicenza), Italien) geb. 1947. Studium der Philologie und Linguistik in Padua und Berkeley. Hauptarbeitsgebiete: Phonetik, Wortstellung und Generative Grammatik. Publikationen: "Studio sperimentale delle caratteristiche elettroacustiche dell'enfasi su sintagmi in italiano" (Mitautor), (in "Fenomeni morfologici e sintattici nell'italiano contemporaneo", SLI, Roma, Bulzoni, 1974); "Studio sperimentale delle caratteristiche elettroacustiche delle vocali toniche e atone in bisillabi italiani" (Mitautor), (in "Studi di fonetica e fonologia", SLI, Roma, Bulzoni, 1976); "101 modi per richiedere" (Mitautor), (in "Aspetti sociolinguistici dell'Italia contemporanea", SLI, Roma, Bulzoni, 1977); "Alcuni aspetti del Tough-Movement nella grammatica relazionale" (Rivista di grammatica generativa, 1977).

FRANÇOIS, Jacques (FB Sprachwissenschaft, Universität Konstanz, Postfach 7733, 7750 Konstanz), geb. 1947 in Paris. Seit 1971 Mitglied der Arbeitsstelle "Documentation et Recherche en Linguistique Allemande, Université de Paris VIII-Vincennes". Z.Zt. Mitarbeiter des Sonderforschungsbereichs 99 ("Vergleichende Lexikologie Deutsch/Französisch) des Fachbereichs Sprachwissenschaft der Universität Konstanz. Hauptarbeitsgebiete: Kontrastive Grammatik, Sprachdidaktik des Deutschen und des Französischen als Zielsprache. Publikationen: Arbeit über eine deutsch-französische Sprachvergleichsgrammatik (Dissertation, 1975); "Zu einer generativen Intersyntagmatik. Die Stellung der Hilfsverben im deutschen, französischen und niederländischen Nebensatz" (Akten des 10. Linguistischen Kolloquiums, Tübingen, Niemeyer, 1976).

GRAFFI, Giorgio (Istituto di Glottologia, Università degli Studi, Strada nuova, 65, 27100 Pavia, Italien) geb. 1949. Studium der Philosophie an der "Scuola Normale Superiore" in Pisa. Wissenschaftlicher Assistent am Lehrstuhl Sprachwissenschaft der Universität Pavia. Hauptarbeitsgebiete: Methodologie und Wissenschaftstheorie der Linguistik, insbesondere der generativen Grammatik, italienische Syntax. Publikationen: verschiedene Aufsätze über Hjelmslev, Chomsky, die Auseinandersetzung zwischen "generativer" und "interpretativer" Semantik.

GRASSL, Wolfgang (Neuholdaugasse 3, 8010 Graz, Österreich) geb. 1955 in Graz. Studium der Anglistik, Philosophie und Sprachwissenschaft in Salzburg und Graz. Hauptarbeitsgebiete: Wissenschaftstheorie und Methodologie der Sprachwissenschaft, logische Semantik. Promoviert z.Zt. am Philosophischen Institut der Universität Graz über das Thema "Empirie und Erklärung in der Linguistik. Eine Studie zur realistischen Interpretation linguistischer Theoriebildung".

HABEL, Christopher (Wilhelmstrasse, 155, 1000 Berlin 20) geb. 1950 in Göttingen. Studium der Mathematik, Linguistik und Philosophie in Göttingen und Osnabrück. Wissenschaftlicher Angestellter am Institut für Angewandte Informatik, Technische Universität Berlin. Hauptarbeitsgebiete: maschinelle Sprachverarbeitung, theoretische Linguistik, Theorie der Sprachvariation. Publikationen: "Ein formales Modell zur Beschreibung von Sprachvariationen" (Akten des 11. Linguistischen Kolloquiums, Tübingen, Niemeyer, 1977); Review-article: "Natürliche Generative Grammatik" (mit S. Kanngiesser)

(Lingua 42, 1977); "Phrasen-Struktur-Grammatiken mit bewerteten Ableitungen" (in Habel/Kanngiesser (eds.) "Sprachdynamik und Sprachstruktur", Tübingen, Niemeyer, 1978).

HAUENSCHILD, Christa (Landfriedstrasse, 12, 6900 Heidelberg) geb. 1949 in Mennighüffen (Kreis Herford). Ausbildung als Dolmetscher für Russisch und Französisch; dann Studium der Sprachwissenschaft und Slawistik in Heidelberg. Mitarbeiter am Teilprojekt A2 "Automatische Übersetzung" des Sonderforschungsbereichs 99 "Linguistik" am Institut für Angewandte Sprachwissenschaft der Universität Heidelberg. Hauptarbeitsgebiete: Erstellung einer computergerechten Beschreibung des Russischen, Übersetzungstheorie. Publikationen: Beiträge zum Forschungsbericht des Teilprojektes an die DFG.

HUCKERT, Edgar (Landfriedstrasse, 12, 6900 Heidelberg) geb. 1950 in Harlingen (Saarland). Studium der Germanistik und Romanistik in Saarbrücken und Toulouse. Angestellter am Teilprojekt A2 "Automatische Übersetzung" des Sonderforschungsbereichs 99 "Linguistik" am Institut für Angewandte Sprachwissenschaft der Universität Heidelberg. Hauptarbeitsgebiete: Erstellung der Programme/Algorithmen für die automatische Übersetzung. Erstellung von Grammatiken für die automatische Übersetzung ins Französische. Publikationen: Beiträge zur automatischen Übersetzung in den Forschungsberichten des SFB 99 und des SFB 100 (Saarbrücken).

LANGE, Klaus-Peter (Westring, 10a, 6740 Landau in der Pfalz) geb. 1938 in Kiel. Studium der klassischen Philologie und der Philosophie in Kiel, Innsbruck und München. Assistenzprofessor am Fachbereich Angewandte Sprachwissenschaft der Universität Mainz. Hauptarbeitsgebiete: Syntax, Semantik, Semiotik. Publikationen: "Theoretiker des literarischen Manierismus - Tesauros und Pellegrinis Lehre von der 'acutezza' - oder von der Macht der Sprache" (München, Fink, 1968). "Strukturen der Syntax" (dt. Übersetzung von N. Chomsky "Syntactic Structures", The Hague, Mouton, 1973); "Kurzgefasste Einführung in die Generative Syntax" (Frankfurt am Main/Bern, Lang, 1975); "Syntax und natürliche Semantik im Deutschen" (erscheint 1978).

MAIER, Robert (Markgräflerstrasse, 8, 6900 Heidelberg) geb. 1944 in Graslitz (Tschechoslowakei). Studium der Mathematik und mathematischen Logik in Giessen und Münster. Wiss. Angestellter am Teilprojekt A2 "Automatische Übersetzung" des Sonderforschungsbereichs 99 "Linguistik" am Institut für Angewandte Sprachwissenschaft der Universität Heidelberg. Hauptarbeitsgebiete: Semantik, semantische Repräsentationssysteme auf logischer Basis. Publikationen: Beiträge in den Forschungsberichten des SFB 99.

NESPOR, Marina (Weteringschans, 233, Amsterdam, The Netherlands) geb. 1949 in Mailand. Studium der Philosophie und der allgemeinen Sprachwissenschaft in Mailand und an der University of North Carolina at Chapel Hill. Assistent am Instituut voor

Italiaanse Taal- en Letterkunde der Universität Amsterdam. Publikationen: Napoli, D. J./Nespor, M., "Negatives in Comparatives" (Language, 52, 1976); Nespor, M./Allen, G.D. "Segment and word durational correlates of syntactic boundaries in Italian" (in R. Rodman (ed.), "Explorations in Language and Linguistics", U.N.C. Occasional Papers in Linguistics, 1).

NEUHAUS, H. Joachim (Westfälische Wilhelms-Universität, Johannisstrasse, 12-20, 4400 Münster) geb. 1945 in Saarbrücken. Professor für Englische Philologie und Linguistik. Hauptarbeitsgebiete: elektronische Sprachforschung, Grammatiktheorie und formale Semantik. Publikationen zur Morphologie und Semantik des Englischen.

PETRACCO SICARDI, Giulia (via F. Romani 15/14, 16122 Genova) geb. 1922 in Turin. Studium in Mailand. Ordentlicher Professor für Sprachwissenschaft an der Universität Genua. Hauptarbeitsgebiete: Onomastik, Germanistik, Romanistik, hochmittelalterliche Latinität in Italien. Publikationen: "Ricerche topografiche e linguistiche sulla Tavola di Polcevera" (Studi genuensi 2, 1958-59); "Toponomastica di Pigna" (Bordighera, Istituto di Studi Liguri, 1962); "Osservazioni sulla lingua dei contratti agrari altomedioevali" (Bollettino Centro Studi Filologici e Linguistici Siciliani 11, 1969); "Contributo linguistico agli studi sulla Tavola Amalfitana" (Torino, Accademia delle Scienze, 1974).

PINKAL, Manfred (Pfarrhausstrasse, 49, 7000 Stuttgart 80) geb. 1949 in Dortmund. Studium der Linguistik, Philosophie, Informatik, Germanistik in Bochum und Stuttgart. Wissenschaftlicher Angestellter am Institut für Linguistik der Universität Stuttgart. Hauptarbeitsgebiete: logische Semantik, Pragmatik. Publikation: "Kontext und Bedeutung" (Dissertation, Tübingen 1977).

POTT, Hartmut (Königsteinerstrasse, 173, 6230 Frankfurt am Main) geb. 1945 in Frankfurt. Studium der Romanistik, Geschichte und Geographie in Frankfurt, Tübingen und Bochum. Studienrat in Frankfurt. Hauptarbeitsgebiete: Generative Semantik, Methodik des Fremdsprachenunterrichts, Aphasiologie. Publikationen: "Studien zur generativen Grammatik" (Hrsg.), (Frankfurt, 1974); "Der Ausdruck der Konzessivität im Französischen"(Bern, Frankfurt, München, 1976); "Möglichkeiten der Anwendung eines modernen Grammatikmodells im fremdsprachlichen Schulunterricht" (Tübingen, 1976); "Linguistische Aspekte zu aphasischen Syndromen" (Tübingen, 1977).

SCHMIDT, Arno (Philippistrasse, 13, 1000 Berlin 19) geb. 1950 in Kirn. Studium der Informatik an der Technischen Universität Berlin. Angestellter am Institut für Angewandte Informatik der Technischen Universität Berlin. Hauptarbeitsgebiet: Maschinelle Sprachverarbeitung.

SCHREYER, Rüdiger H. (Jülicher Strasse, 1, 5102 Würselen-Broichweiden) geb. 1941 in Frankenholz. Studium der Anglistik, Amerikanistik und Germanistik in Saarbrücken. Akademischer Oberrat am Institut für Anglistik der RWTH Aachen. Hauptarbeitsgebiete: Geschichte der Sprachwissenschaft, Sprachtheorie, Stratifikationsgrammatik, Semantik. Publikationen: " Untersuchungen zur Sprachauffassung Dr. Johnsons"; "Stratifikationsgrammatik. Eine Einführung"(Tübingen, Niemeyer, 1977).

SCHWEPPE, Heinz (Margaretenstrasse, 33, 1000 Berlin 45) geb. 1947. Studium der Mathematik an der TU Hannover und an der Universität Zürich. Wissenschaftlicher Assistent am Institut für Angewandte Informatik der T.U. Berlin. Hauptarbeitsgebiete: Datenbanken, Informationssysteme, maschinelle Sprachverarbeitung.

von STECHOW, Arnim (Stephansplatz, 18, 7750 Konstanz), geb. 1941 in Berlin. Studium der Allgemeinen Sprachwissenschaft, Romanistik, Philosophie, Mathematischen Logik und Geschichte in Bonn,Grenoble und Münster. Professor an der Universität Konstanz. Hauptarbeitsgebiete: Semantik und Syntax. Publikationen: (mit A. Kratzer und E.Pause): "Einführung in Theorie und Anwendung der generativen Syntax" (Frankfurt, 1973). "Zur formalen Syntax und Semantik des Deutschen" (Arbeitspapiere des SFB 99, Konstanz, 1975); (mit A. Kratzer): "Äusserungssituation und Bedeutung" (LiLi, Heft 23:24 "Psycholinguistik", 1977).

TEKAVČIĆ, Pavao (Berislavićeva, 12, 4100 Zagreb, Jugoslawien) geb. 1931 in Zagreb. Studium des Französischen, Italienischen, und Lateinischen an der Universität Zagreb. Ordentlicher Professor für italienische Linguistik an der Universität Zagreb. Hauptarbeitsgebiete: Italienische Linguistik, Romanistik, Vulgärlatein. Publikationen: "Grammatica storica dell'italiano" (in drei Bänden, Bologna, Il Mulino, 1972); "Uvod u vulgarni latinitet (s izborom tekstova)", (Universität Zagreb, 1970); "Abozzo del sistema morfosintattico del soprasilvano odierno" (Studia Romanica et Anglica Zagrabiensia, 33-36, 37).

van de VELDE, Marc (Pr. P. Fredericqstraat, 42, 9000 Gent, Belgien) geb. 1944 in Zottegem. Studium der Germanistik an der Rijksuniversiteit Gent. Dozent am Moger Instituut voor vertalers en tolken, Gent. Hauptarbeitsgebiete: Wortstellung (kontrastiv deutsch-niederländisch); Syntax.

VIETHEN, Heinz W. (Institut für Anglistik III der RWTH Aachen, Kármánstrasse, 17-19, 5100 Aachen) geb. 1947 in Düren. Studium der Anglistik, Germanistik, Phonetik, Allgemeinen Sprachwissenschaft in Köln und Manchester. Wissenschaftlicher Assistent am Institut für Anglistik III der RWTH Aachen. Hauptarbeitsgebiete: Phonetik/Phonologie, Syntax/Semantik, Sprachtests. Publikationen: Rezension: "Visser, F.Th. (1969): An Historical Syntax of the English Language. Part III, First Half: Syntactical Units with Two Verbs" (in Anglia 91, 1973); Herausgeber (mit W.D. Bald und K. Sprengel) der Akten des 11. Linguistischen Kolloquiums, Aachen 1976 (Tübingen, Niemeyer, 1977); "All, any, each, every und generisches a" (Akten des 11. Linguistischen Kolloquiums, Tübingen, Niemeyer, 1977); "Untersuchungen zum englischen Present Perfect und zum Generischen" (Frankfurt am Main/Bern, Lang, im Druck).

WIRRER, Jan (Fakultät für Linguistik und Literaturwissenschaft, Universität Bielefeld, 48 Bielefeld) geb. 1944 in Würzburg. Studium der Germanistik, Anglistik und Philosophie an den Universitäten Hamburg, Freiburg, Kiel, Tübingen. Wissenschaftlicher Assistent an der Universität Bielefeld. Hauptarbeitsgebiete: Natürliche Textverarbeitung, Didaktik des Deutschunterrichts, empirische Literaturwissenschaft. Publikationen: Beiträge in Kindt/Schmidt (eds.),"Interpretationsanalyse". Verschiedene Review-articles. Weitere Artikel in verschiedenen Zeitschriften und Sammelbänden.

VERZEICHNIS DER HERAUSGEBER

CONTE, Maria-Elisabeth (Istituto di Storia della lingua italiana, Università, Strada nuova, 65, 27100 Pavia), geb. 1935 in Soest in Westfalen. Studium der Anglistik und Romanistik in Frankfurt, Münster, Besançon und Freiburg im Breisgau. Professore incaricato der Semiotik an der Universität Pavia. Hauptarbeitsgebiete: Semantik, Pragmatik, Texttheorie, Geschichte der Sprachwissenschaft. Publikationen: "L'aggettivo in italiano" (in "Storia linguistica dell'Italia nel Novecento", SLI, Roma, Bulzoni, 1973); "Wilhelm von Humboldt nella linguistica contemporanea" (Lingua e stile 8, 1973); "Semantische und pragmatische Ansätze in der Sprachtheorie Wilhelm von Humboldts" (in H. Parret (ed.), "History of Linguistic Thought and Contemporary Linguistics", Berlin, de Gruyter, 1976); M.-E. Conte (ed.), "La linguistica testuale" (Milano, Feltrinelli, 1977).

GIACALONE RAMAT, Anna (via Franchi, 3, 27100 Pavia) geb. 1937 in Forlí. Studium der klassischen Philologie, Germanistik und Linguistik in Florenz und Würzburg. Ordentlicher Professor für Sprachwissenschaft an der Universität Pavia. Hauptarbeitsgebiete: historische Sprachwissenschaft, Germanistik, Soziolinguistik, Sprachvariation. Publikationen: verschiedene Beiträge zur Indogermanistik und zur Sprachgeschichte des Lateinischen und des Gotischen. Soziolinguistik der deutschen Sprachinseln am Monte Rosa (in Vorbereitung).

RAMAT, PAOLO (Istituto di Glottologia, Università, 27100 Pavia) geb. 1936 in Florenz. Studium der Indogermanistik, Germanistik und des Sanskrit in Florenz, Würzburg und Saarbrücken. Ordentlicher Professor für Sprachwissenschaft an der Universität Pavia. Hauptarbeitsgebiete: Germanistik, Typologie, Allgemeine Sprachwissenschaft. Publikationen: "Grammatica dell'antico sassone" (Milano, Mursia, 1969); P. Ramat (ed.), "F. Engels, Storia e lingua dei Germani" (Roma, Editori Riuniti, 1975); "Das Friesische. Eine sprachliche und kulturgeschichtliche Einführung" (Innsbruck, 1976); P. Ramat (ed.), "La tipologia linguistica" (Bologna, Il Mulino, 1976); P. Ramat (ed.), "Manuale di linguistica germanica" (in Vorbereitung). Organisator einer Tagung der SLI (=Società di Linguistica Italiana) über "Rekonstruktion in der Linguistik" ("La ricostruzione in linguistica", SLI, Roma Bulzoni, 1977).

FSC
www.fsc.org
MIX
Papier | Fördert
gute Waldnutzung
FSC® C083411

Kolibri 360 GmbH
Ferdinand-Jühlke-Straße 7
99095 Erfurt, Deutschland
produktsicherheit@kolibri360.de